Mehran Mousapoor

Industrial Water Tube Boiler Design

Also of Interest

Optimized Equipment Lubrication.
Conventional Lube, Oil Mist Technology and Full Standby Protection
Heinz Bloch, 2021
ISBN 978-3-11-074934-2, e-ISBN 978-3-11-074944-1

Wind Energy & System Engineering
Dexin He and GuiYong Yu, 2022
ISBN 978-3-11-040143-1, e-ISBN 978-3-11-040153-0

Compressor Technology Advances.
Beyond 2020
Hurlel Elliott and Heinz Bloch, 2021
ISBN 978-3-11-067873-4, e-ISBN 978-3-11-067876-5

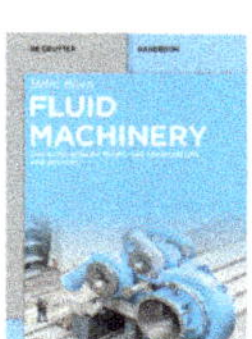

Fluid Machinery.
Life Extension of Pumps, Gas Compressors and Drivers
Heinz Bloch, 2020
ISBN 978-3-11-067413-2, e-ISBN 978-3-11-067415-6

Mehran Mousapoor

Industrial Water Tube Boiler Design

Formulas in Practice

DE GRUYTER

Author
Mr Mehran Mousapoor
+98 (912) 4210204
m.mousapoor@gmail.com

ISBN 978-3-11-075661-6
e-ISBN (PDF) 978-3-11-075708-8
e-ISBN (EPUB) 978-3-11-075713-2

Library of Congress Control Number: 2021943129

Bibliographic information published by the Deutsche Nationalbibliothek
The Deutsche Nationalbibliothek lists this publication in the Deutsche Nationalbibliografie; detailed bibliographic data are available on the Internet at http://dnb.dnb.de.

Cover image: Mehran Mousapoor
Typesetting: Integra Software Services Pvt. Ltd.
Printing and binding: CPI books GmbH, Leck

www.degruyter.com

To my parents Mohammad and Tahereh,
for inspiring me to write this book
To my wife Mona,
for strong support, without her this project would not have been done
and to my son Hafez,
without his inspiration, this book would not have been completed.
It is for the manner all of them have tolerated my lack of colorful presence during the past three years.

132T/hr, 11.5 barg, 200 °C Water tube boiler, Bidboland II Refinery, Behbahan, Iran

Contents

1 Introduction

This guideline gives direction as to how to design a water tube D-type boiler. This design guideline can help engineers understand the basic design of a boiler with an appropriate dimension, materials and heat of combustion. Good execution of a boiler is affected by the greatest heat absorbed and least heat misfortune. The design of the boiler may be impacted by variables, counting process requirements, financial issues and safety. All vital parameters and equations used within this guideline are clarified, which help the readers understand the meaning of parameters or the term utilized. The theory area clarifies how to calculate sizing and determination of a boiler. This guideline makes a difference by the readers to get almost the heat balance concept. The application of the boiler's theory with the illustrations will make the engineers understand boiler's basics and perform the actual design of boilers [5].

1.1 Boiler basics

A boiler is a closed vessel in which liquid (for the most part, water) is heated. The fluid does not necessarily boil. The heated or vaporized fluid exits the boiler for use in different processes or heating applications [1]. Water tube boiler is a shape of the boiler in which steam is produced by circulating water through tubes exposed to the source of heat.

Boilers utilize a heat source, ordinarily combustion of a fossil fuel, to heat water to generate hot water or steam. Boilers nowadays burn fuel gas and oil as well as solid fuels and proceed to play a bigger role in industries.

Boilers are separated by their arrangement, dimension and the quality of the steam or hot water delivered. Boiler's dimension is most regularly measured by the fuel input in million Btu per hour (MMBtu/h). Size may be measured in pounds of steam per hour (pph). Output may be measured in horsepower. One boiler horsepower = 33,475 Btu/h evaporation capacity or about 34.5 pph. Since huge boilers are regularly used to create power, it may moreover be valuable to relate boiler size to power output in electric producing applications. Utilizing ordinary boiler and producing efficiencies, 100 MMBtu/h heat input is increased to almost 10 MW electric output. Hot water boilers, for the most part, heat water to 250 °F or less at pressures of 250 pounds per square inch (psig) or less. Huge numbers of small low-pressure steam boilers (<10 MMBtu/h) have been utilized at small factories or operate in support of larger manufacturing processes. Low-pressure steam boilers for the most part deliver saturated steam at temperatures of 350–400 °F at pressures between 125 and 250 psig. The bigger steam usage is for industry, power generation and area heating. The main steam consuming industries are refineries, petrochemical, power plants, paper, food industries and metal factories. Large boilers generate high-pressure steam and may be

https://doi.org/10.1515/9783110757088-001

rated at 250–10,000 MMBtu/h. High-pressure boilers can generate steam temperatures above 700 °F and pressures till 3,000 psig [2].

1.2 Boiler types

Boilers can be characterized by the arrangement of heat transfer surfaces – either fire tube or water tube – and by the combustion system. The appropriate arrangement is determined by suitable fuel, steam conditions and capacity. Figure 1.1 shows the capacity ranges between fire tube and water tube boilers on a heat input basis [3].

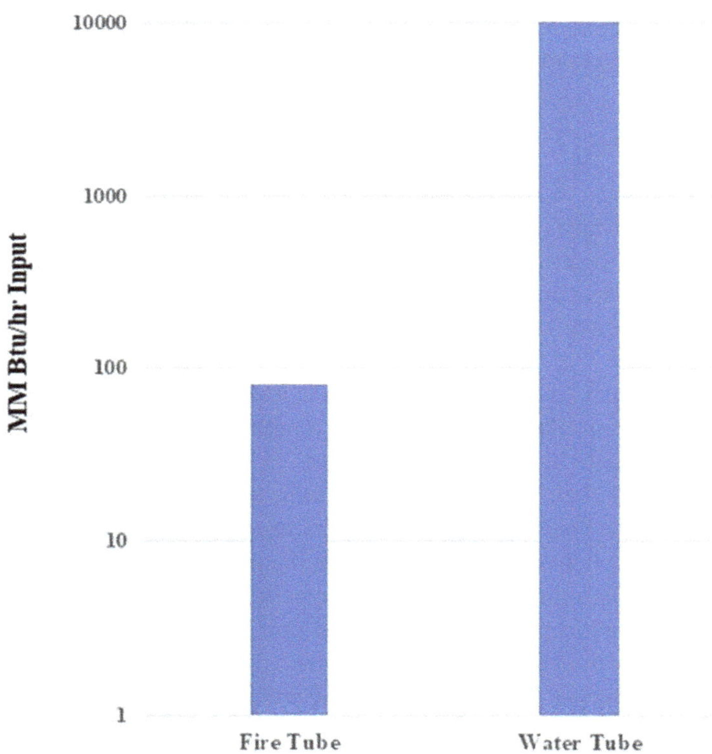

Fig. 1.1: Water tube versus fire tube boiler [3].

1.3 Boiler arrangement

There are two types of boilers. In a fire tube boiler, the water is in the main part of the boiler, and the combustion gases pass through metal tubes. Heat is transferred to the water by conduction from the fire tube(s) to the surrounding water. Increasing the number of "passes" that combustion gases make through the boiler upgrades heat extraction. The simplicity and low cost of fire tube boilers are their point of interest, nearly all fire tube boilers burn oil, natural gas or both. The mixing of water in a huge chamber makes a fire tube boiler well suited to generate hot water or low-pressure steam. For high-pressure (>200 psig) or high-capacity (>10 MMBtu) applications, fire tube boilers (Fig. 1.2) are not desirable because of pressure vessel failure, which in a water tube boiler is just can be failure of a single tube [3].

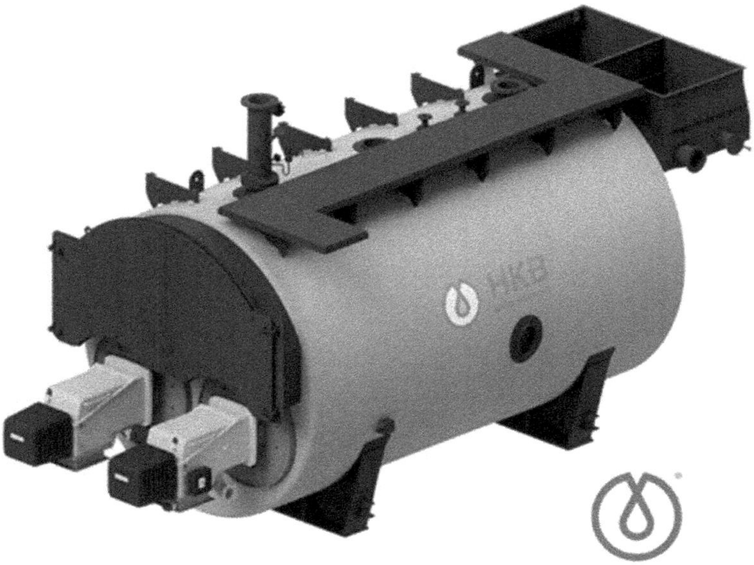

Fig. 1.2: Fire tube boiler, HKB Boiler Solution.

In water tube boilers (Fig. 1.3), the fuel burns in a furnace and the exhaust gases flow surround metal tubes. Heat transfer to the water tubes is achieved by radiation, conduction and convection. The water tubes are welded together to form the shape of the combustion chamber in a "waterwall." Water tube boilers can generate steam at high temperatures and pressures, and these boilers are more complex and expensive than fire tube boilers [3].

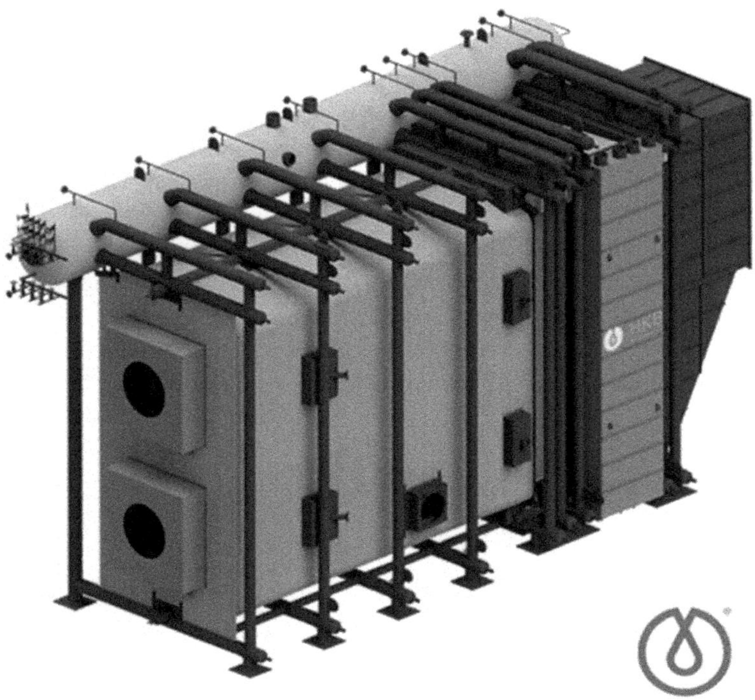

Fig. 1.3: Water tube boiler, HKB Boiler Solution.

1.4 Boiler designing sequence

Sequences for designing a water tube boiler in this book will be as follows:
1) Calculating heat duty of a boiler is based on subtracting outlet steam enthalpy from inlet steam enthalpy and multiplying outlet steam flow rate.
2) Calculate the required fuel and combustion air for burning mixture of air and fuel.
3) Calculate the heat input rate by generating burners and then selecting burner quantity and size from burner manufacturer catalog.
4) Calculate the width and height of a furnace, which is approximate of flame diameter and height.
5) Calculate the flame temperature, density and flue gas volumetric flow rate to find the required volume of furnace. For this matter, velocity of flue gas will be taken from boiler manufacturer handbook. Finally, we can calculate the height of furnace. Please note that this height, width and length are preliminary and need to recalculate the end of our calculations.

6) At this point, some items such as furnace flue gas draft pressure drop, furnace flue gas exit temperature, heat flux and tube metal temperature need to be controlled.
7) Calculate the boiler design pressure.
8) Designing of a superheater package includes the design of superheater tube size, thickness, length, tube rows number and tube rows deep number. Please note that the mentioned item will be calculating preliminary and by calculating heat duty and optimization, corrected amount will be obtained.
9) Calculate the outlet steam and flue gas temperature from superheater package.
10) Calculate the steam and mud drum diameter and thickness.
11) Calculate the height between drums center, and then bank tubes average length.
12) Designing of a bank tube package by predicting its heat duty, and then predicting tube rows number, tube rows deep number, heat flux, steam drum outlet steam temperature and bank tube flue gas draft pressure drops. By iteration and checking limits, our prediction comes to reality, and final data will be obtained.
13) Designing of an economizer package by predicting heat duty to obtain final size and specification, which is same as that of superheater and boiler bank packages.
14) In all mentioned sequence draft, pressure drops, steam pressure drops, velocities and dimension of each package should be checked and even total package dimensioned should be controlled.
15) Till now we assumed the circulation ratio which is related to the dimension and pressure of boiler and we continued our calculation. But at this moment we should calculate and control the circulation ratio.
16) At this point, stack and safety valves will be sized.
17) We are calculating our boiler's efficiency at the next step. This item can show us our design was good or need to be revised. If it will be okay, then we can go through drum hold-up time and number of tubes and each part and package's weight.
18) All reports can be seen at the end of this book.

2 Input data

In this book, we try to design a water tube boiler with the following descriptions:

- Boiler type: D-type, natural circulation, outdoor installation
- Required high pressure steam: 60 T/h = 132,277.2 lb/h
- Required steam pressure: 42.83 kg/cm^2 g = 609.15 psig
- Required steam temperature: 420 °C = 788 °F
- Feed water temperature: 110 °C = 230 °F
- MCR (maximum continuous rating) condition: 25%, 50%, 75%, 100%
- Blowdown rate: 3%
- Burner quantity: 2 (each burner covers 75% of boiler MCR)
- Stack: 20 m height, installed separated on the ground
- Total dissolved solids in water: 1 ppm
- Site elevation: 4 m above sea level
- Wind velocity: 33 m/s
- Assumed stack flue gas inlet temperature: 160 °C = 320 °F
- Fuel: see natural gas specification shown in Tab. 2.1.

Tab. 2.1: Natural gas specification.

	%mol
CH_4	85.9284
C_2H_6	6.2249
C_3H_8	1.5041
i-C_4H_{10}	0.2087
n-C_4H_{10}	0.2676
i-C_5H_{12}	0.0963
n-C_5H_{12}	0.0642
C_6H_{14}+	0.0642
CO_2	1.3007
H_2S (ppm)	4
N_2	4.3409
Total	100

https://doi.org/10.1515/9783110757088-002

Tab. 2.1 (continued)

	%mol
LHV (kcal/kg)	10,971
HHV (kcal/kg)	12,144

Note: It should be considered that all calculation at first step will be estimation and need to recalculate after detailed design.
LHV, lower heating value; HHV, higher heating value.

3 Boiler heat duty

The heat duty can be explained as the amount of heat that is transferred from a hot side to the cold side in a unit of time. First, we should calculate the boiler heat duty as follows:

$$Q_{\text{to steam}} = h_{\text{outlet steam}} - h_{\text{inlet water}}$$

$$h_{\text{outlet steam}} = 777.85\ \text{kcal/kg}$$

$$h_{\text{inlet water}} = 110\ \text{kcal/kg}$$

$$\downarrow$$

$$\mathbf{Q_{\text{to steam}} = 667.85\ kcal/kg}$$

$$\downarrow$$

$$Q_{\text{duty}} = Q_{\text{to steam}} \times m'_{\text{boiler at 100\% MCR}}$$

$$Q_{\text{duty}} = 667.85\ \text{kcal/kg} \times 60\ \text{t/hr} = 40,071.4\ \text{M kcal/hr}$$

$$\mathbf{Q_{\text{duty}} = 40,071.4\ M\ kcal/h = 159\ \ MM\ Btu/hr = 46.603\ MW}$$

https://doi.org/10.1515/9783110757088-003

4 Required fuel

Therefore, after calculating our boiler heat duty, we should find basic boiler data such as required fuel and combustion air. For this part, we should have data from our fuel type, and for this book, the data which we are considering natural gas as our fuel (Tab. 4.1) is common in all petrochemical, refinery or power plants.

Tab. 4.1: Natural gas specification.

	%mol	Molecular weight factor	Molecular weight
CH_4	85.9284	16.041	13.784
C_2H_6	6.2249	30.067	1.872
C_3H_8	1.5041	44.092	0.663
i-C_4H_{10}	0.2087	58.118	0.121
n-C_4H_{10}	0.2676	58.118	0.156
i-C_5H_{12}	0.0963	72.144	0.069
n-C_5H_{12}	0.0642	72.144	0.046
C_6H_{14}+	0.0642	86.169	0.055
CO_2	1.3007	44.01	0.572
H_2S (ppm)	4		0.000
N_2	4.3409	28.016	1.216
Total	100		**18.555**
LHV (kcal/kg)			**10,971**
HHV (kcal/kg)			**12,144**

LHV, lower heat value; HHV, higher heat value.

$$m'_{\text{fuel}} = (Q_{\text{duty}} \times 1.05)/\text{LHV}$$

$$m'_{\text{fuel}} = (40{,}071.4 \times 1{,}000 \times 1.05)/10{,}971$$

$$\boldsymbol{m'_{\text{fuel}} = 3{,}835.1\ \text{kg/h}}$$

https://doi.org/10.1515/9783110757088-004

5 Forced draft fan discharge mass flow

Forced draft fan discharge mass flow will calculate by multiplying air density and actual combustion air flow rate. Sequence will be given as follows:

- Barometric pressure
- Saturation pressure at actual temperature
- Moisture in air
- Total required air
- Excess air
- Total required combustion air
- Air density
- Standard combustion air flow
- Actual combustion air flow
- Forced draft fan discharge mass flow

5.1 Barometric pressure

Barometric pressure is the weight of the air over us. The Earth's atmosphere contains air and it is moderately light and its weight as gravity pulls the air atoms.

The nominal barometric pressure on the Earth is concurred to be 101.325 kPa absolute (1,013.25 mbar absolute or 14.696 psi absolute), which suggests that there is around 1.03 kgf/cm^2 (14.7 lbf/in^2) regularly on the Earth's surface caused by the weight of the air. In practice, the barometric pressure exceptionally once in a while is precisely that nominal value, because it is changing all the time and shifts at different areas.

The barometric pressure also changes based on altitude. When you go higher, you will find the smaller barometric pressure, which makes sense that if one of you moves to a higher elevation, there is less air on top of you.

The air at higher heights moreover includes fewer particles, making it lighter than it would be at a lower height. The gravity moreover decreases at these heights. Due to these reasons, the barometric pressure is smaller at higher elevations (Tab. 5.1) [4].

Tab. 5.1: Ambient conditions.

Temperature	20	°C
Minimum temperature	–20	°C
Relative humidity	60	%
Altitude	4	m

https://doi.org/10.1515/9783110757088-005

$$P = \frac{101,325 \times (1 - 2.25577 \times 10^{-5} \times H)^{5.25588}}{6,894.76}$$

where P is barometric pressure of air (psia) and H is the altitude above sea level (ft)

$$P = \frac{101,325 \times (1 - 2.25577 \times 10^{-5} \times 13.12)^{5.25588}}{6,894.76}$$

$$\boldsymbol{P} = \mathbf{14.69\ psia}$$

5.2 Saturation pressure at actual temperature

At moist air, the saturation pressure of water vapor is different with the temperature of mixture air vapor and can be expressed as follows [5]:

$$P_{\mathrm{ws}} = \frac{e^{\left(77.3450 + 0.0057 \times T - \frac{7,235}{T}\right)}}{(T^{8.2}) \times 6,894.76}$$

where P_{ws} is water vapour saturation pressure (pisa), e is the constant 2.718 and T is dry bulb temperature of moist air (K)

$$P_{\mathrm{ws}} = \frac{e^{\left(77.3450 + 0.0057 \times (20 + 273.15) - \frac{7,235}{(20 + 273.15)}\right)}}{(20 + 273.15)^{8.2}}$$

$$\boldsymbol{P_{\mathrm{ws}}} = \mathbf{0.3381(psia)}$$

5.3 Moisture in air

For calculating nonluminous heat transfer, we ought to know the full amount of water vapor in flue gases, a portion of which comes from combustion air.

Moreover, if atmospheric air compresses, the saturated vapor pressure of water will be higher, and if the air is cooled underneath the comparing water dew point temperature, water can condense.

It is critical to know the sum of water vapor in air or flue gas because the sum of moisture in air or gas flxes the water dew point temperature [5]:

$$M = 0.622 \times \frac{\mathrm{RH\%} \times \mathrm{SVP}}{(P_{\mathrm{barometric}} - \mathrm{RH\%} \times \mathrm{SVP})}$$

where M is moisture in air, RH is relative humidity and SVP is the saturated vapor pressure (psia)

$$M = 0.622 \times \frac{0.6 \times 0.4861}{(14.69 - 0.6 \times 0.3381)}$$

$$\mathbf{M = 0.009\ lb/lb\ dry\ air}$$

5.4 Total required air

Every fuel such as natural gas, coal or oil requires a certain percentage of stoichiometric air per MMBtu fired (on higher heating value basis) [5]. Total air required for each fuel will be used in Tab. 5.2.

Tab. 5.2: Calculation from combustion constant table.

	Molecular weight	Air required for combustion [5]	LB air required for combustion
CH_4	13.784	17.265	14.835
C_2H_6	1.872	16.119	1.003
C_3H_8	0.663	15.703	0.236
i-C_4H_{10}	0.121	15.487	0.0323
n-C_4H_{10}	0.156	15.487	0.041
i-C_5H_{12}	0.069	15.353	0.014
n-C_5H_{12}	0.046	15.353	0.009
C_6H_{14}+	0.055	15.266	0.009
CO_2	0.572	0	0
H_2S (ppm)	0.000		0
N_2	1.216	0	0
	18.555		**16.183**
	Net required air (lb/lb dry air)		16.183
	Moisture in dry air (lb/lb dry air)		0.009
	Total required air (lb/lb dry air)		**16.192**

5.5 Excess air

Theoretically, each amount of fresh air can be blended by a fixed amount of fuel and that will be burnt as ideal combustion. In this regard, each fuel will have been properly burnt and consumed all oxygen in the air. Efficiency of combustion will be high when no excess air exists.

Actually, ideal combustion is not possible. Theoretically in complete combustion, a few portions of fresh air would give insufficient oxygen, and the amount of carbon tends to be converted into carbon monoxide rather than carbon dioxide. Air shortage will lead to dangerous levels of carbon monoxide being formed and will then produce smoke.

Hence, it is normal to adjust the combustion with a level of excess air as safety margin [6].

One practical guide to find excess air is to use from burner manufacturer catalog such as Pillard company. Table 5.3 shows us the oxygen rate and then we can find the excess air in different boiler loads:

$$\text{Excess air\%} = 1.1 \times \frac{O_2}{20.8 - O_2}$$

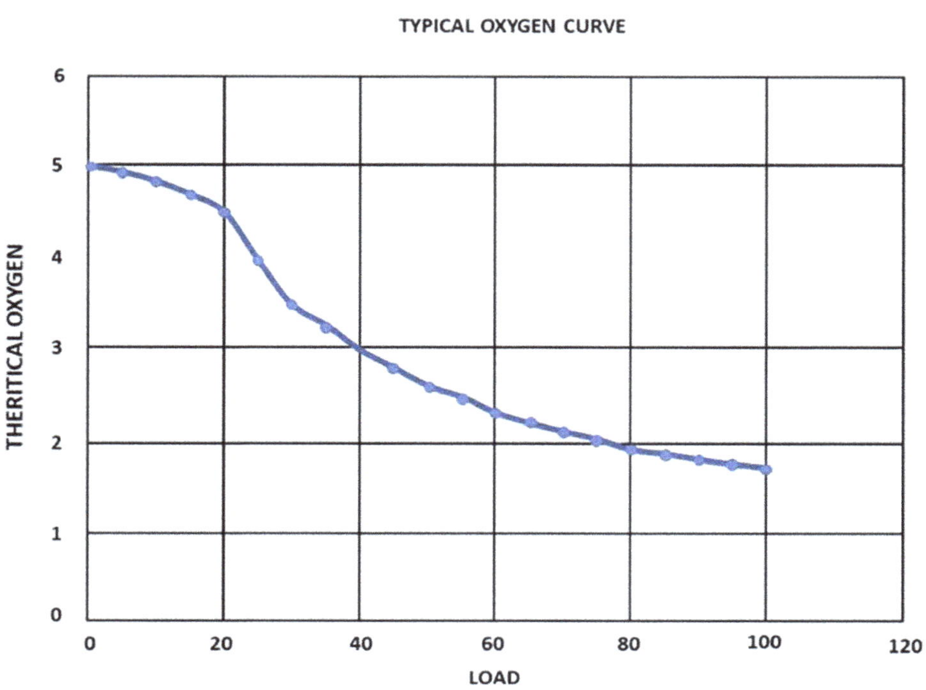

Fig. 5.1: Typical oxygen curve by Fives Pillard.

Tab. 5.3: Excess air calculation.

Boiler MCR %	**100%**	**75%**	**50%**	**25%**
Oxygen rate	1.8	2	2.6	4
Calculated excess air %	10.42	11.70	15.71	26.19
Selected excess air %	17	21	28	44

Note: It should be considered that all calculation at first step will be estimation and need to be recalculated after detailed design.

5.6 Total required combustion air

For calculating the performance of blower, major problems occur in specifying actual cubic feet per minute (ACFM) from standard cubic feet per minute (SCFM), and in accurately converting from one to other one. Some engineers use SCFM and some ACFM.

SCFM is regularly designed by flow in terms of pressure, temperature and humidity [7].

These corrections should be made to assure that complete amount of oxygen will be provided to function properly:

$$W'_{\text{total combustion air}}(\text{kg/h}) = \text{Total air required}(\text{lb/lb dry air}) \times m'_{\text{fuel}}\ (\text{kg/h}) \times (1 + \text{excess air})$$

$$W'_{\text{total combustion air}} = 16.192 \times 3,835.1 \times 1.17$$

$$W'_{\textbf{total combustion air}} = \mathbf{72,655\ (kg/h)}$$

5.7 Air density

The density of any gas can be estimated from

$$\rho_g = 492 \times \text{MW} \times \frac{P}{359 \times (460 + T) \times 14.7}$$

where ρ_g is the gas density (lb/cu.ft), P is gas pressure (psia), T is gas temperature (°F) and MW is molecular weight of gas

$$\rho_{\text{air}} = 492 \times 29 \times \frac{14.69}{359 \times (460 + (1.8 \times 20 + 32)) \times 14.7}$$

$$\rho_{\text{air}} = \mathbf{0.0752\,lb/cu.ft}$$

5.8 Standard combustion air flow

In the same mass flow rate, variety in standard temperature will lead to a significant volumetric variety [5]:

$$V_{\text{s}} = \frac{W'_{\text{s}}}{\rho_{\text{air}} \times 60}$$

$$V_{\text{s}} = \frac{72,655(\text{kg/h}) \times 2.20462}{0.0752 \times 60}$$

$$\mathbf{V_s = 35,492.67(SCFM)}$$

5.9 Actual combustion air flow

The volume of flowing gas will be defined as ACFM, which considers its temperature and pressure. In the event that the system was moving a gas at precisely the "standard" condition, then ACFM would equal SCFM. Tragically, the most important change between these two definitions is the pressure.

The positive pressure or a vacuum must be produced to flow a gas. It is compressed when positive pressure is connected to a standard cubic foot of gas. It expands when a vacuum is connected to a standard cubic foot of gas. After pressurizing or rarefying, volume of gas will be the "actual" volume [8]

$$\frac{V_{\text{a}}}{V_{\text{s}}} = \frac{P_{\text{s}}}{P_{\text{a}}} \times \frac{T_{\text{a}}}{T_{\text{s}}}$$

where V_{a} is actual combustion air volume S CFM, V_{s} is standard combustion air volume SCFM, P_{s} is standard air pressure at sea level p sia, P_{a} is air pressure at actual level p sia, T_{s} is standard ambient air temperature (R) and T_{a} is actual ambient air temperature (R)

$$V_{\text{a}} = \frac{14.7}{14.69} \times \frac{(1.8 \times 20 + 491.67)}{520} \times 35,492.67$$

$$\mathbf{V_a = 36,048.20\ (SCFM)}$$

5.10 Forced draft fan discharge mass flow

$$W'_{\mathrm{a}} = 60 \times \rho_{\mathrm{air}} \times V_{\mathrm{a}}$$

$$W'_{\mathrm{a}} = 60 \times 0.0752 \times 36,048.20$$

$$\mathbf{W'_{a} = 162,683.59(lb/h) = 73,792.12(kg/h)}$$

6 FD: Fan outlet duct design

For designing width, height and length of an outlet duct, it must be first consuming fan discharge pressure, location from boiler and duct roots, width and height first and after calculating pressure loss, need to recalculate it to find optimum duct size and reach the minimum air pressure loss.

With air flow rate and permissible velocity in duct, we can find the duct's minimum area and then by assuming the width or height, we can estimate another, and the total duct size must be going through minimum pressure loss:

$$W'_{\mathrm{a}} = 162{,}683.59(\mathrm{lb/h})$$

Force draft air velocity range inside the boiler = 1,500–3,600 fpm [41]

Selected velocity = 3,600 fpm

MW = 29.24

$T = 68\ °\mathrm{F}$

$P_{\mathrm{barometric}} = 14.69$ psia

Assumed boiler draft pressure loss = 1.2 psi

$P_{\mathrm{discharge}} = 15.89$ psia

$$\rho_{\mathrm{air}} = 492 \times 29.24 \times \frac{15.89}{359 \times (460 + 68) \times 14.7}$$

$$\boldsymbol{\rho_{\mathrm{air}} = 0.082\ \ (\mathrm{lb/ft^3})}$$

Volumetric flow rate of air is given by

$$V'_{\mathrm{air}} = \frac{162{,}683.59}{0.082}$$

$$\mathbf{V'_{air} = 1{,}983{,}452.7(cu.ft/h)}$$

Air duct minimum area is given by

$$A_{\mathrm{min}} = \frac{1{,}983{,}452.7 \times 60}{3{,}600}$$

https://doi.org/10.1515/9783110757088-006

$$\boldsymbol{A_{\min} = 9.18\text{ft}^2}$$

From fan manufacturer catalog for these ranges, preliminary width and height can be found and here we assumed width (a) as 23 in and then we calculate height (b) and will finally check the pressure loss:

$$b = \frac{A_{\min}}{a}$$

$$b = \frac{9.18 \times 144}{24}$$

$$\boldsymbol{b = 55.10\text{ in}}$$

↓

The rectangular duct equivalent diameter will be defined by

$$d_i = 1.3 \times \frac{(a \times b)^{0.625}}{(a+b)^{0.25}}$$

$$d_i = 1.3 \times \frac{(24 \times 55.1)^{0.625}}{(24+55.1)^{0.25}}$$

$$\boldsymbol{d_i = 38.93\text{ in}}$$

7 FD.Fan outlet duct pressure loss

The viscosity of air can be deducted from Tab. 7.1:

$$\mu_{air} = 0.0459 \ \text{lb/fth}$$

Tab. 7.1: Air viscosity [5].

Temp. (°F)	Viscosity (lb/ft h)
100	0.0459
200	0.0520
400	0.062
600	0.0772
800	0.0806
1,000	0.0884
1,200	0.0957
1,400	0.1027
1,600	0.1100
1,800	0.1512

The Reynolds number of air is given by

$$\text{Re} = 15.2 \times \frac{W'}{\mu \times d_i}$$

$$\text{Re} = 15.2 \times \frac{162{,}683.59}{0.0459 \times 38.93}$$

$$\mathbf{Re = 1{,}384{,}006.64}$$

The friction factor of turbulent flow of air in ducts is given by

$$\text{f} = \frac{0.316}{\text{Re}^{0.25}}$$

$$\text{f} = \frac{0.316}{1{,}384{,}006.64^{0.25}}$$

$$\mathbf{f = 0.009}$$

For calculating duct pressure loss, we should know about the duct arrangement and equivalent length. Here we consider one sudden enlargement and contraction

https://doi.org/10.1515/9783110757088-007

as 31.863 ft and then we find our equivalent length as 61.863 ft, and the duct pressure loss is calculated as follows:

$$\Delta P_{\text{air}} = 9.3 \times 10^{-6} \times f \times W'^2 \times \frac{L_e}{\rho \times d_i^5}$$

$$\Delta P_{\text{air}} = 9.3 \times 10^{-6} \times 0.009 \times 162,683.59^2 \times \frac{61.863}{0.082 \times 38.93^5}$$

$$\boldsymbol{\Delta P_{\text{air}} = 0.1914 \text{ in } H_2O = 4.86 \text{ mm } H_2O}$$

This pressure loss is reasonable and common in boilers, so this design of duct is acceptable as the boiler pressure loss is calculated in detail in earlier chapters which need to coming back recalculate again.

8 Furnace width and length

Furnace width and length will be calculated by selecting the burner capacity and size, and flame length and diameter. Sequence will be as follows:

- Heat input to furnace
- Burner capacity selection
- Furnace width and length

8.1 Heat input to furnace

For designing furnace and calculating the size of furnace, we must calculate the required heat input to furnace to have mentioned heat duty and use from burner manufacturer catalog for capacity and flame dimension and then we can find preliminary furnace dimensions. In our design, we consider that each burner should work till 75% of boiler's maximum continuous rating:

$$Q'_{\text{burner}} = \text{LHV} \times m'_{\text{fuel}} \times 110\%$$

$$Q'_{\text{burner}} = 10{,}971\,(\text{kcal/kg}) \times 3{,}835.1\,(\text{kg/h}) \times \frac{3.96832\left(\text{Btu/h}\big/\text{kcal/h}\right)}{10^6} \times 110\%$$

$$Q'_{\text{burners}} = 184\,(\text{MM Btu/h}) = 53.83\,\text{MW}$$

$$\mathbf{Q'_{burner} = 184 \times 0.75 = 138\,(MM\,Btu/h)}$$

8.2 Burner capacity selection

Burner capacity is the range between the minimum and maximum BTUs that can be generated by a burner with a stable flame and acceptable combustion. In other words, burner capacity will specify the ranges of minimum and maximum firing. When selecting a burner, make sure that any models you consider have proper capacity for the defined process. Small capacity will lead to a smaller load and that will prevent from reaching the required temperature.

From one burner manufacturer such as Pillard company for ultra-low NO_x burners catalog (Bulletin 4211) for 138 MMBtu/h capacity, we can use from model no. 4211-48 in which its flame length and diameter are 25 and 6.3 feet, respectively (Tab. 8.1).

https://doi.org/10.1515/9783110757088-008

Tab. 8.1: Flame dimension [39].

Input at 10% XSA (million Btu/h)	**Flame length (ft)**	**Flame diameter (ft)**
4211-21	12.5	2.8
4211-25	13	3
4211-27	13.5	3
4211-33	14.5	3.3
4211-38	15	3.5
4211-44	15.5	3.8
4211-48	16.5	4
4211-54	17	4.3
4211-62	18	4.5
4211-74	20	4.8
4211-86	21.5	5
4211-98	22.5	5.5
4211-111	23.5	5.8
4211-116	24	6
4211-124	24.5	6
4211-140	**25**	**6.3**
4211-163	27	6.5
4211-182	29	7
4211-219	31	7.5
4211-256	33	7.8
4211-292	35	8

8.3 Furnace width and length

Furnace minimum length can be estimated as 40 inch more than the length before lance from experience, and also furnace width can be estimated as 20 inch more than the flame width for safety conditions. Distance between burners will be 10 inch less than the flame width. Please note that according to burner designer's point, flame overlap will increase the total flame length by 18 inch.

Fig. 8.1: Rampco boiler, 14th Iranian Oil Exhibition, Tehran, Iran.

Minimum furnace length before lance = 25 × 12 + 60 = 360 in
Minimum furnace width = 6.3 × 12 + 20 = 96 in
Minimum furnace overall length = 360 + 40 = 400 in
Distance between burners = 6.3 × 12 − 10 = 65.6 in

Please note all the abovementioned minimum length and width are first guess, and after designing boiler and circulation ratio calculation, they should be optimum and the following are the optimum length after few circles of calculations:
Furnace length before lance = 400 in
Furnace width = 100 in
Furnace overall length = 440 in
Furnace turning lane = 440–400 = 40 in

For calculating the furnace volume, first we should find furnace height and for that subject, first, we should calculate it from the flow and density of flue gas. So, we need to find the flue gas temperature and then the average flue gas temperature as shown in Fig. 8.2.

Fig. 8.2: Water tube boiler (132 T/h, 42 barg, 390 °C), Basra Petrochemical, Basra, Iraq.

9 Furnace height

Furnace height will be calculated by the following sequence:
- Flue gas flow rate
- Flue gas molecular weight
- Actual flame temperature
- Flue gas temperature inside furnace
- Furnace pressure
- Flue gas inlet density
- Flue gas volumetric flow rate
- Furnace height

9.1 Flue gas flow rate

Flue gas flow rate calculation is given in Tab. 9.1.

Tab. 9.1: Flue gas flow rate calculation.

Items	%mol	From combustible tables			LB flue gas product CO_2	LB flue gas product H_2O	LB flue gas product N_2
		Flue gas product CO_2	Flue gas product H_2O	Flue gas product N_2			
CH_4	85.9284	2.744	2.246	13.275	2.358	1.930	11.407
C_2H_6	6.2249	2.927	1.798	12.394	0.182	0.112	0.772
C_3H_8	1.5041	2.994	1.634	12.074	0.045	0.025	0.182
i-C_4H_{10}	0.2087	3.029	1.55	11.908	0.006	0.003	0.025
n-C_4H_{10}	0.2676	3.029	1.55	11.908	0.008	0.004	0.032
i-C_5H_{12}	0.0963	3.05	1.498	11.805	0.003	0.001	0.011
n-C_5H_{12}	0.0642	3.05	1.498	11.805	0.002	0.001	0.008
C_6H_{14}+	0.0642	3.064	1.464	11.738	0.002	0.001	0.008
CO_2	1.3007	0	0	0	0.000	0.000	0.000
H_2S (ppm)	4				0.000	0.000	0.000
N_2	4.3409	0	0	0	0.000	0.000	0.000
	100				**2.61**	**2.08**	**12.44**

https://doi.org/10.1515/9783110757088-009

Tab. 9.1 (continued)

Items	%mol	From combustible tables			LB flue gas product CO_2	LB flue gas product H_2O	LB flue gas product N_2
		Flue gas product CO_2	Flue gas product H_2O	Flue gas product N_2			
			Total flue gas product (lb)			**17.13**	lb flue gas/lb fuel
			Moisture in dry air			0.009	lb/lb dry air
			Total flue gas product (lb)			**17.27**	lb flue gas/lb fuel

$$W'_{\text{flue gas}}(\text{lb/h}) = \text{Total fuel required }(\text{lb/lb}_{\text{flue gas}}) \times m'_{\text{fuel}}(\text{lb/h}) \times (1 + \text{excess air})$$

$$W'_{\text{flue gas}} = 17.27 \times 3,835.1 \times 2.20462 \times 1.17$$

$$\mathbf{W'_{\text{flue gas}} = 170,818.92(lb/h)}$$

9.2 Flue gas molecular weight

$$\text{MW}_{\text{flue gas}} = \text{MW}_{CO_2} \times \text{lb}_{CO_2} + \text{MW}_{H_2O} \times \text{lb}_{H_2O} + \text{MW}_{N_2} \times \text{lb}_{N_2}$$

$$\text{MW}_{\text{flue gas}} = 44.01 \times 2.61 + 18 \times 2.08 + 28.016 \times 12.44$$

$$\mathbf{MW_{\text{flue gas}} = 29.23}$$

9.3 Actual flame temperature

Maximum temperature of fuel and air combustion is adiabatic combustion temperature. In any case, due to separation and radiation losses, this maximum is never achieved.

For specifying the final temperature, several equations must be solved. Practically, the actual combustion temperature is 3–5% lower than the adiabatic combustion temperature [5].

By an energy balance, it can be shown as follows:

$$t_c = \frac{\text{LHV} + A \times \alpha \times \text{HHV} \times C_{\text{pa}} \times \left(\frac{t_a - 80}{10^6}\right)}{(1 - \%\text{ash}/100 + A \times \alpha \times \text{HHV}/10^6) \times C_{\text{pg}}}$$

LHV, HHV are lower and higher calorific values of fuel, Btu/lb; A is the theoretical air required per million Btu fired, lb; α = excess air factor = 1 + (E%/100); t_a, t_c are temaeratues of aier and combustion, °F; C_pa, C_pg are specific heats of air and products of combustion, Btu/lb.F.

Tab. 9.2: Combustion constant (A) for fuels [5].

Blast furnace gas	575
Bagasse	650
Carbon monoxide gas	670
Refinery and oil gas	720
Natural gas	730
Furnace oil and lignite	745–750
Bituminous coals	760
Anthracite	780
Coke	800

$$t_c = \frac{19{,}747.8 + 730 \times 1.17 \times 21{,}860.81 \times 0.24 \times (68 - 80)/10^6}{(1 - \%/100 + 730 \times 1.19 \times 21{,}860.81/10^6) \times 0.36}$$

$$t_\text{c} = 2{,}780.98\ ^\circ\text{F}$$

Actual flame temperature = adiabatic flame temperature × 0.95

$$\mathbf{t_{ca} = 2{,}641.93\ ^\circ F}$$

9.4 Flue gas temperature inside furnace

Flue gas temperature (t_g) is specified in numerous ways; some authors define it as the exit gas temperature. Some others show it as the mean of the theoretical flame temperature. In any case, experiences appear that better agreement between measured and calculated values wins when [5]

$$t_\text{g} = t_\text{c} + 300 \text{ to } 400\ ^\circ\text{F}$$

$$t_g = 2,641.93 + 300$$

$$\mathbf{t_g = 2,941.93\ ^\circ F = 1,616.63^\circ C = 3,401.93\ ^\circ R}$$

9.5 Furnace pressure

Furnace heat transfer may be a complicated circumstantial, and a single equation or correlation cannot be endorsed for sizing of furnaces. In general, it will be defined as an energy balance between fluid–gas and steam–water mixtures [5].

From burner manufacturer catalog, we could find flame length and width and also estimate furnace length and width

$$P_{\text{furnace}} = P_{\text{fan outlet}} - P_{\text{loss air duct}} - P_{\text{loss burner}}$$

$$P_{\text{furnace}} = 15.89 - 0.1942 \times 0.0361273 - 250 \times 0.00142233$$

$$\mathbf{P_{furnace} = 15.53\ psia}$$

9.6 Flue gas inlet density

$$\rho_g = 492 \times \text{MW} \times \frac{P_{\text{furnace}}}{359 \times (460 + T) \times 14.7}$$

$$\rho_g = 492 \times 29 \times \frac{15.53}{359 \times (460 + 2,941.93) \times 14.7}$$

$$\boldsymbol{\rho_g} = \mathbf{0.0123\ lb/cu.ft}$$

9.7 Flue gas volumetric flow rate

$$V'_{\text{flue gas}} = \frac{m'_{\text{flue gas}}}{\rho_{\text{flue gas}}}$$

$$V'_{\text{flue gas}} = \frac{170,818.92}{0.0123}$$

$$\mathbf{V'_{flue\ gas} = 13,843,237\ ft^3/h}$$

9.8 Furnace height

Furnace height can be found out as follows (Fig. 9.1):

Fig. 9.1: Water tube boiler (132 T/h, 42 barg, 390 °C), Basra Petrochemical, Basra, Iraq.

$$\text{Selected velocity inside furnace} = 3{,}000\ \text{fpm}$$

$$A_{\text{min}} = \frac{V'_{\text{flue gas}}}{V_{\text{flue gas}}} \times 1.1$$

$$A_{\text{min}} = \frac{13{,}843{,}237}{3{,}000 \times 60} \times 1.1$$

$$\boldsymbol{A_{\text{min}} = 84.59\ \text{ft}^2}$$

$$H_{\text{min}} = \frac{A_{\text{min}}}{W}$$

$$H_{\text{min}} = \frac{84.59 \times 144}{100}$$

$$\boldsymbol{H_{\text{min}} = 134\ \text{in}}$$

By considering the distance between burners, furnace height can be calculated by the following equilibrium:

- burners with 50% MCR capacity will be estimated by
  - Burner quantity × flame diameter – 20 + 13 × burner quantity
- burners with 75% MCR capacity will be estimated by
  - Burner quantity × flame diameter – 20 + 13 in

$$H_{\text{furnace}} = \text{flame width} \times \text{burner QTY} - 20 + 13$$

$$H_{\text{furnace}} = 6.3 \times 12 \times 2 - 20 + 13$$

$$\boldsymbol{H_{\text{furnace}} = 144 \text{ in}}$$

From the above calculation, furnace height is bigger than the calculation by velocity method and furnace height can be considered as 144 inch.

10 Furnace volume

$$V = L \times W \times H$$

$$V = \frac{440 \times 100 \times 144}{12^3}$$

$$\boldsymbol{V = 3,671.75 \mathrm{ft}^3}$$

https://doi.org/10.1515/9783110757088-010

11 Furnace exit temperature

Furnace exit temperature will be calculated by the following sequence:

- Furnace volumetric heat release
- Furnace heat surface area
- Furnace surface heat release
- Beam length
- Flue gas carbon dioxide vapor pressure
- Flue gas water vapor pressure
- Gas emissivity by carbon and water method
- Gas emissivity by Hottel's method
- Furnace exit temperature

11.1 Furnace volumetric heat release

Furnace volumetric heat release should be lower than 450,000 kcal/m^3 h.

$$\text{Volumetric heat release} = \frac{\text{Heat input to furnace}}{V_{\text{furnace}}}$$

$$\text{Volumetric heat release} = \frac{184\ \text{MM Btu/h}}{3{,}671.75\ \text{ft}^3}$$

$$\textbf{Volumetric heat release} = \mathbf{45{,}473.25\ Btu/ft^3\ h\ =\ 404{,}402.51\ kcal/m^3\ h}$$

11.2 Furnace heat surface area

$$A_{\text{furnace}} = \frac{2 \times W \times H + 2 \times W \times L_{\text{before lance}} + 2 \times H \times L_{\text{before lance}} + (L - L_{\text{before lance}}) \times H}{144}$$

$$A_{\text{furnace}} = \frac{2 \times 100 \times 144 + 2 \times 100 \times 400 + 2 \times 144 \times 400 + (440 - 400) \times 144}{144}$$

$$\boldsymbol{A_{\text{furnace}}} = \mathbf{1{,}597\ ft^2}$$

11.3 Furnace surface heat release

Primary function of a furnace is to supply adequate space for fuel particles to burn completely and to cool down the flue gas to a temperature through which the convective heating will operate and this depends on the type of utilized firing method [9].

https://doi.org/10.1515/9783110757088-011

The average furnace heat absorption rate based on the projected water-cooled surface shall not exceed 200,000 kcal/m^2 h and the maximum peak absorption rate in the burner zone shall not exceed 300,000 kcal/m^2 h

$$\text{Surface heat release} = \frac{\text{Heat input to furnace}}{\text{Furnace heat surface area}}$$

$$\text{Surface heat release} = \frac{\text{LHV} \times m'_{\text{fuel}}}{A_{\text{furnace}}}$$

$$\text{Surface heat release} = \frac{10{,}971(\text{kcal/kg}) \times 3{,}835.1(\text{kg/h}) \times \dfrac{3.96832(\text{Btu/h})/(\text{kcal/h})}{10^6}}{1{,}597(\text{ft}^2)}$$

$$\textbf{Surface heat release} = \mathbf{104{,}550.31\ Btu/ft\ h = 283{,}589\ kcal/m^2\ h}$$

11.4 Beam length

L may be a characteristic measurement which depends on the shape of the wall in an area. *L* is estimated around 3.4–3.6 times the volume of the space divided by the surface zone of the heat-receiving surface. For a depth of dimensions, *L*, *W* and *H* are given by [5]

$$L = \frac{3.4 \times LWH}{2 \times (LW + LH + WH)}$$

$$L = \frac{1.7}{1/L + 1/W + 1/H}$$

$$\boldsymbol{L = 87.47\ \text{in}}$$

11.5 Flue gas carbon dioxide vapor pressure

$$P_{v\text{CO}_2} = \frac{\text{lb}_{\text{CO}_2}}{\text{lb}_{\text{flue gas}}}$$

$$P_{v\text{CO}_2} = \frac{2.61}{17.27}$$

$$\boldsymbol{P_{v\text{CO}_2} = 0.15}$$

11.6 Flue gas water vapor pressure

$$P_{vH_2O} = \frac{lb_{H_2O} + moisture_{air}}{lb_{flue\ gas}}$$

$$P_{vH_2O} = \frac{2.08\ +\ 0.009}{17.27}$$

$$\mathbf{P_{vH_2O} = 0.12}$$

11.7 Gas emissivity by carbon and water method

Hottel calculated the emissivity pattern of gases, and gas emissivity can be predicted by gas temperature, partial pressure of gases and beam length [5]:

- ε_g is given by

$$\varepsilon_g = \varepsilon_c + \eta\varepsilon_w - \Delta\varepsilon$$

ε_g can be estimated using Figs. 11.1–11.4, which give $\varepsilon_c, \eta, \varepsilon_w$ and $\Delta\varepsilon$, respectively. It can be estimated by assuming radiation effects of SO_2 as similar to CO_2. Consequently, partial pressures of CO_2 and SO_2 can be added, and Fig. 11.1 is used to get ε_c:

$$P_{vCO_2} \times L_{beam} = \frac{0.15 \times 87.47}{12} = 1.1$$

$$P_{vH_2O} \times L_{beam} = \frac{0.12 \times 87.47}{12} = 0.88$$

$$\frac{P_{vH_2O} + P}{2} = \frac{(0.12 + 1)}{2} = 0.56$$

$$\frac{P_{vH_2O}}{P_{vH_2O} + P_{vCO_2}} = \frac{0.12}{(0.12 + 0.15)} = 0.44$$

$$(P_{vH_2O} + P_{vCO_2}) \times L_{beam} = (0.12 + 0.15) \times 87.47 = 1.98$$

From lower figure

$$\varepsilon_g = 0.098 + 0.11 \times 1 - 0.048$$

$$\varepsilon_g = 0.16$$

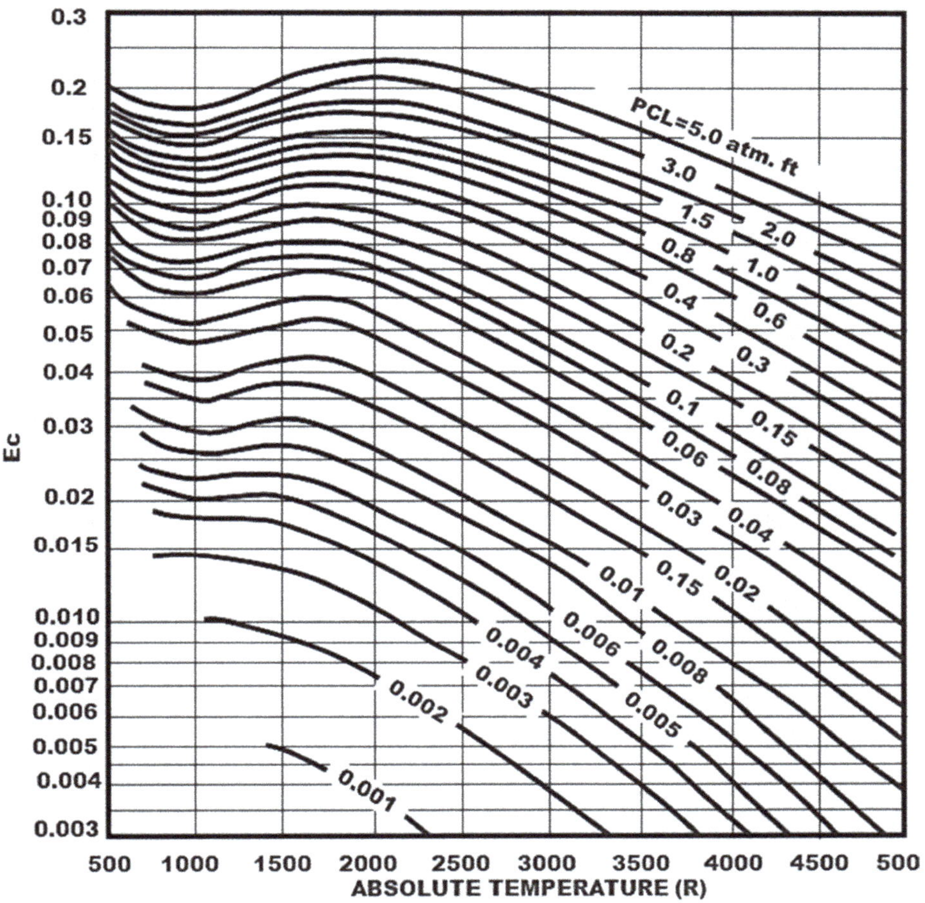

Fig. 11.1: Emissivity of carbon dioxide [5].

11.8 Gas emissivity by Hottel's method

$$\varepsilon_g = 0.9 \times \left(1 - e^{-KL_{\text{beam}}}\right)$$

$$K = \frac{(0.8 + 1.6 \times P_{v\text{H}_2\text{O}}) \times 1 - 0.38\text{T}_g/1,000)}{\sqrt{(P_{v\text{H}_2\text{O}} + P_{v\text{CO}_2})} \times L_{\text{beam}}} \times (P_{v\text{H}_2\text{O}} + P_{v\text{CO}_2})$$

$$K = \frac{(0.8 + 1.6 \times 0.15) \times (1 - 0.38 \times (1{,}616.63 + 273.5)/1{,}000)}{\sqrt{(0.15 + 0.12) \times 87.47 \times 0.0254}} \times (0.15 + 0.12)$$

$$\mathbf{K = 0.097}$$

$$\varepsilon_g = 0.9 \times (1 - e^{-(0.097 \times 87.25)})$$

$$\boldsymbol{\varepsilon_g = 0.175}$$

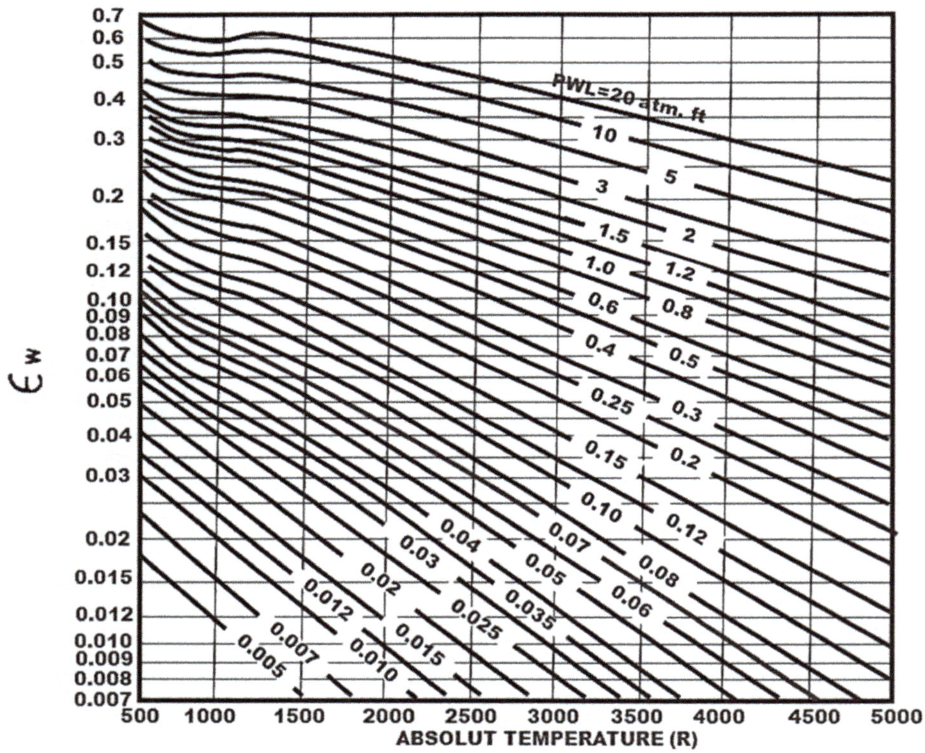

Fig. 11.2: Emissivity of water vapor [5].

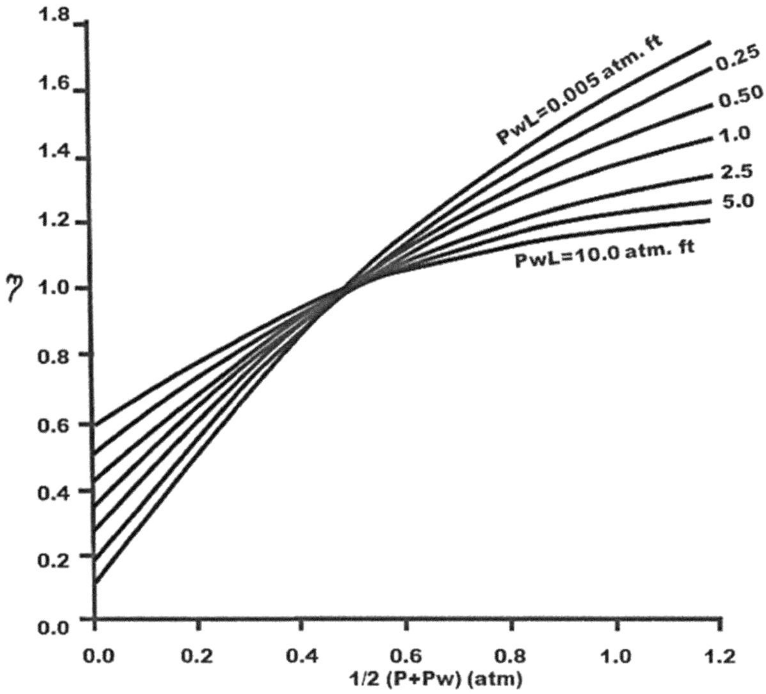

Fig. 11.3: Correction factor for emissivity of water vapor [5].

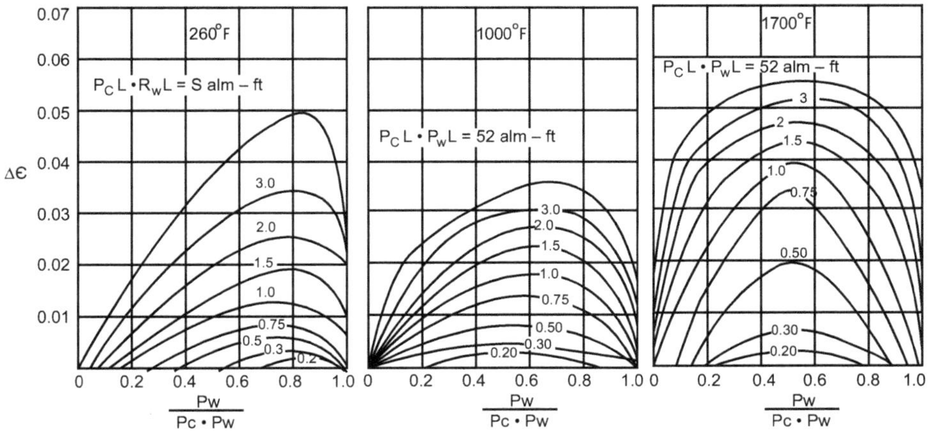

Fig. 11.4: Correction term due to presence of water and carbon dioxide [5].

11.9 Furnace exit temperature

Heat transfer in a boiler furnace is a transcendent radiation which is incomplete due to the luminous part of the flame and nonluminous gases. A common inexact expression can be composed for furnace absorption using a vitality approach (Fig. 11.5) [5]:

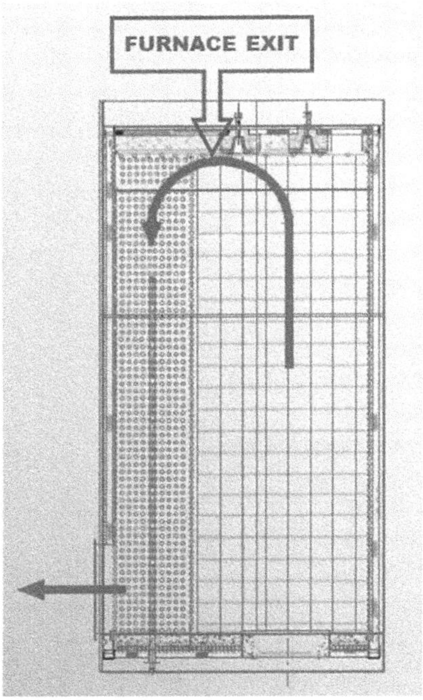

Fig. 11.5: Furnace exit.

$$Q_F = A_P \varepsilon_g \sigma \left(T_g{}^4 - T_o{}^4\right) = W_f \mathrm{LHV} - W_g h_e$$

Q_F is furnace heat absorbtion (Btu/h),
A_P is furnace heat surface $\left(\mathrm{ft}^2\right)$,
σ is the Stefan Boltzmann constant $= 0.173 \times 10^{-8}$,
T_{exit} is furnace exit gas temperature (R),
ε_g is emissivity of gas at T_g,
T_o is absolute temperature of tube surface (R),
W_f is fuel gas flow rate (lb/h),
W_g is flue gas flow rate (lb/h) and
$h_{average}$ is enthalpy at flue gas average temperature (Btu/lb).

In the above equation, we assumed furnace exit temperature and will use from average temperature to find enthalpy. Then by solving both sides (left and right), T_g will be found which should be equal to what we assumed before.

First, we assumed that T_{exit} is equal to 1,296.1 °C:

$$T_{average} = \frac{(T_g + T_{exit})}{2}$$

$$T_{average} = \frac{(2{,}941.93 + (1.8 \times 1{,}296.1 + 32))}{2}$$

$$\boldsymbol{T_{average} = 2{,}666.2\ ^{\circ}F = 1{,}463.5\ ^{\circ}C}$$

Enthalpy is shown in Tab. 11.1:

$$h_{average} = 450.3\,\text{kcal/kg} = 810.54\,\text{Btu/lb}$$

The tube's outer temperature is calculated by deducting the temperature drop across the gas film and metal conductivity and adding temperature drop across the steam film tube's inside temperature.

For the following calculation, the tube's outer temperature will be assumed to be 646 °F which is done by iterations and it should be noted that 100 °F difference in tube's outer temperature has maximum 10 °F effluent on furnace exit temperature:

$$A_P \varepsilon_g \sigma (T_{exit}{}^4 - T_o{}^4) = W_f \text{LHV} - W_g h_{average}$$

Tab. 11.1: Enthalpy of combustion product (kcal/kg) [5].

Temp (°C)	Natural gas	Fuel oil
1,900	606.9	583.9
1,800	570.4	549.2
1,700	534.3	514.8
1,600	498.4	480.6
1,400	428.0	413.1
1,200	359.3	347.0
1,000	292.7	282.8
800	228.2	220.5

$$1,597 \times 0.16 \times 0.173 \times 10^{-8} \times \left(T_{\text{exit}}{}^4 - 646^4\right) = 3,835.1 \times 2.20462 \times 19,747.8 - 170,818.92 \times 810.54$$

$$T_{\text{exit}} = 2,390.5\ ^\circ\text{F} = 1,310.28\ ^\circ\text{C}$$

So, it is found that our assumed temperature is near to the calculated temperature.

12 Combustion nonluminous heat transfer coefficient

In heat transfer equipment, gases transfer energy at high temperatures to fluid which is inside the tubes, such as boilers, fired heaters and superheaters, and nonluminous heat transfer is very important. The produced water vapor, carbon dioxide and sulfur dioxide during combustion of fossil fuels or triatomic gases contribute to radiation.

Radiation compatibility between gases and surroundings (e.g., a wall or tube bundle or a cavity) can be composed as follows [5]:

$$\frac{Q}{A} = \sigma\left(\varepsilon_{\mathrm{g}} T_{\mathrm{g}}{}^{4} - \alpha_{\mathrm{g}} T_{\mathrm{o}}{}^{4}\right)$$

where Q is heat transfer in furnace (Btu/h),
A is furnace heat surface (ft^2),
σ is the Stefan Boltzmann constant $= 0.173 \times 10^{-8}$,
ε_{g} is emissivity of gas at T_{g},
T_{g} is the absolute temperature of gas, R
α_{g} is absorptivity at T_{o},
T_{o} is the absolute temperature of tube surface, R.

In spite of the fact that it is alluring to calculate heat flux, it is repetitive to estimate α_{g} at temperature T_{o}. We can use the following simplified equation by considering the reality that $T_{\mathrm{o}}{}^{4}$ will be very smaller than $T_{\mathrm{g}}{}^{4}$:

$$\frac{Q}{A} = \sigma\left(\varepsilon_{\mathrm{g}} T_{\mathrm{g}}{}^{4} - \alpha_{\mathrm{g}} T_{\mathrm{o}}{}^{4}\right) = h_{\mathrm{N}}\left(T_{\mathrm{g}} - T_{\mathrm{o}}\right)$$

The nonluminous heat transfer coefficient h_{N} can be composed as follows:

$$h_{\mathrm{N}} = \frac{\sigma \varepsilon_{\mathrm{g}}\left(T_{\mathrm{g}}{}^{4} - T_{\mathrm{o}}{}^{4}\right)}{\left(T_{\mathrm{g}} - T_{\mathrm{o}}\right)}$$

$$\downarrow$$

$$h_{\mathrm{N}} = \frac{0.173 \times 10^{-8} \times 0.16 \times \left(2{,}390.5^{4} - 646^{4}\right)}{\left(2{,}390.5 - 646\right)}$$

$$\boldsymbol{h_{\mathrm{N}} = 11.25\ \mathrm{Btu/ft^2\,h\,F}}$$

https://doi.org/10.1515/9783110757088-012

13 Combustion convection heat transfer coefficient

Combustion's convection heat transfer coefficient will be calculated by the following sequence:

- Flue gas product properties
- Flue gas product gas mass velocity
- Flue gas product Reynolds number
- Flue gas product Prandtl number
- Flue gas product Nusselt number
- Combustion convection heat transfer coefficient

13.1 Flue gas product properties

Flue gas properties at average temperature can be found from Tab. 13.1:

$$T_{\text{average}} = \frac{(2,941.93 + 2,390.5)}{2}$$

$$T_{\text{average}} = 2,666.2\,°\text{F} = 1,463.5\,°\text{C}$$

Tab. 13.1: Flue gas product properties [5].

Temp (°F)	C_p	M	K
2,600	0.35105	0.136	0.0622
2,500	0.347975	0.1329	0.06035
2,400	0.3449	0.1298	0.0585
2,300	0.341825	0.1267	0.05665
2,200	0.33875	0.1236	0.0548

$$\mu = 0.136 \text{ lb/ft h}$$

$$C_p = 0.35105 \text{ Btu/ft F}$$

$$K = 0.0622 \text{ Btu/ft h F}$$

13.2 Flue gas product gas mass velocity

Gas mass velocity is mass flow rate over a unit range opposite to the direction of the velocity vector and can be found as follows:

https://doi.org/10.1515/9783110757088-013

$$G = \frac{m'_{\text{flue gas}}}{A_{\text{furnace}}}$$

$$\boldsymbol{G = \frac{170,818.92}{1,597} = 106.96\,\text{lb}/\text{ft}^2\,\text{h}}$$

13.3 Flue gas product Reynolds number

The Reynolds number has wide applications, extending from fluid stream in a pipe to the entry of air over an airplane wing. It is utilized to foresee the movement from laminar to turbulent stream and is utilized within the scaling of comparative but different-sized stream circumstances, such as between an aircraft demonstrate in a wind tunnel and the full-size form. The forecast of the onset of turbulence and the capacity to calculate scaling impacts can be utilized to assist foresee fluid behavior on a bigger scale, such as in nearby or worldwide air or water movement and subsequently the related meteorological and climatological impacts and can be found as follows [11]:

$$\text{Re} = \frac{Gd_{\text{Tube}}}{12\mu}$$

$$\text{Re} = \frac{106.96 \times 2}{12 \times 0.136}$$

$$\mathbf{Re = 131.08}$$

13.4 Flue gas product Prandtl number

The Prandtl number (Pr) is a dimensionless number, named after the German physicist Ludwig Prandtl, characterized as the proportion of momentum diffusivity to thermal diffusivity and can be found as follows: [12]

$$\text{Pr} = \frac{\mu C_p}{K}$$

$$\text{Pr} = \frac{0.136 \times 0.35105}{0.0622}$$

$$\mathbf{Pr = 0.767}$$

13.5 Flue gas product Nusselt number

The Nusselt number (Nu) could be explained at a boundary in a fluid as the proportion of convective to conductive heat transfer and is a dimensionless number. Convection incorporated fluid motion and conduction. The conduction for a theoretically motionless fluid is measured beneath the same conditions as the convection.

A Nusselt number represents heat transfer by pure conduction. A value between 1 and 10 is characteristic of slug stream or laminar stream. Bigger Nusselt number inside turbulent flow ordinarily in the 100–1,000 range corresponds to more dynamic convection. The Nusselt number is named after Wilhelm Nusselt and can be found as follows [13]:

$$\mathrm{NU} = 0.33 \times \mathrm{Re}^{0.6} \times \mathrm{Pr}^{0.33}$$

$$\mathrm{NU} = 0.33 \times 131.08^{0.6} \times 0.767^{0.33}$$

$$\mathbf{NU = 5.638}$$

13.6 Combustion convection heat transfer coefficient

In thermodynamics and mechanics, heat transfer coefficient or film coefficient or film effectiveness is the proportionality that is steady between the heat flux and the thermodynamic driving force for the stream of heat (i.e., the temperature difference, ΔT), and the heat transfer coefficient is the complementary of thermal insulance and is given as follows [14]:

$$h_c = \frac{12K \times \mathrm{NU}}{d_{\mathrm{Tube}}}$$

$$\mathrm{h_c} = \frac{12 \times 0.0622 \times 5.638}{2}$$

$$\boldsymbol{h_c = 2.10\ \mathrm{Btu/ft^2\,h\,F}}$$

14 Combustion outside heat transfer coefficient

$$h_o = h_c + h_N$$

$$h_o = 2.10 + 11.25$$

$$\boldsymbol{h_o = 13.36\ \mathbf{Btu/ft^2\ h\ F}}$$

https://doi.org/10.1515/9783110757088-014

15 Average tube metal temperature

Tube's outer temperature will be calculated by deducting temperature drop across the gas film and metal conductivity and adding temperature drop across the steam film tube's inside temperature [5].

Sequence for finding the average tube metal temperature is given as follows:

- Drum circulation water mass flow rate
- Drum each tube circulation water
- Inside gas film's resistance
- Metal resistance
- Outside gas film's resistance
- Heat flux
- Temperature drop across the gas film
- Temperature drop across the tube metal
- Temperature drop across the steam film
- Average tube metal temperature

15.1 Drum circulation water mass flow rate

First, we should find each tube's water mass flow which is equal to the boiler water mass flow (output steam mass flow rate + blowdown rate) divided by the boiler's total tube numbers:

$$m'_{\text{drum water}} = m'_{\text{output steam}} \times (1 + \%_{\text{blow down rate}})$$

↓

$$m'_{\text{output steam}} = 60 \times 1,000 \times 2.20462 = 132,277.2 \ \text{lb/h}$$

$$\%_{\text{blow down rate}} = 3\%$$

↓

$$m'_{\text{drum water}} = 132,277.2 \times (1 + 0.03)$$

$$\mathbf{m'_{\text{drum water}} = 136,245.516\ lb/h}$$

https://doi.org/10.1515/9783110757088-015

15.2 Drum each tube circulation water

Boiler's total tube number is equal to the boiler bank tube's total tube number plus furnace tube number:

$$N_{\text{boiler tube}} = N_{\text{total bank tubes}} + N_{\text{furnace tube}} + 2 \times N_{\text{rows deep}_{\text{bank tube}}}$$

Later, there are all formula to calculate boiler tube number, but here we use from them and continue our sequence to find each tube's mass flow. Please note that by one iteration all the tube quantity will be optimized:

$$N_{\text{boiler tube}} = 976 + 413 + 2 \times 14$$

$$\mathbf{N_{\text{boiler tube}} = 1,417No.}$$

↓

Total circulated water inside the drum is related to boiler circulation ratio which is described later but now we consider our circulation ratio as 40, and boiler's total circulated water is equal to the drum input water multiplied by circulation ratio:

$$m'_{\text{drum circulation water}} = m'_{\text{drum water}} \times \text{CR}$$

$$m'_{\text{drum circulation water}} = 136,245.516 \times 40$$

$$\mathbf{m'_{\text{drum circulation water}} = 5,499,820.64 \;\; lb/h}$$

↓

$$m'_{\text{each tube circulation water}} = \frac{m'_{\text{drum circulation water}}}{N_{\text{boiler tube}}}$$

$$m'_{\text{each tube circulation water}} = \frac{5,499,820.64}{1,417}$$

$$\mathbf{m'_{\text{each tube circulation water}} = 3,846.03 \, lb/h}$$

15.3 Inside gas film resistance

$$R_{\text{Inside gas film}} = \frac{1}{h_{\text{tube inside coefficient}}}$$

↓

$$h_{\text{tube inside coefficient}} = \frac{2.44 \times m'_{\text{each tube circulation water}}{}^{0.8} \times C}{(d_{\text{tube}} - t_{\text{tube}})^{1.8}}$$

↓

C factor is given in Tab. 15.1 by drum saturated pressure and temperature, in which the drum saturated pressure is superheater output pressure plus superheater and boiler bank and another pressure loss. Here, total pressure loss can be considered as 1–6 barg and after final designing it can be optimized by iteration:

$$P_{\text{sat.}} = P_{\text{super heater outlet}} + \Delta P_{\text{assumption}}$$

$$P_{\text{sat.}} = 42 + 2.22 = 44.22\,\text{barg}$$

$$\boldsymbol{P_{\text{sat}} = 643.69\,\text{psia}}$$

$$\boldsymbol{T_{\text{sat.}} = 493.8\,^{\circ}\text{F}}$$

↓

$$\boldsymbol{C = 0.36}$$

Tab. 15.1: Factor *C* for steam [5].

Temperature (°F)	Drum saturation pressure (psia)				
	100	200	500	1,000	2,000
400	0.2716	0.3059			
500	0.2737	0.2909	0.3595		
600	0.2813	0.2896	0.3228	0.413	
700	0.2917	0.2965	0.3161	0.3586	0.5206
800	0.3050	0.3090	0.3206	0.3453	0.4214
900	0.3161	0.3197	0.3277	0.3477	0.3946
1,000	0.3276	0.3302	0.3392	0.3531	0.386

↓

$$h_{\text{tube inside coefficient}} = \frac{2.44 \times 3{,}846.03^{0.8} \times 0.36}{(2-0.105)^{1.8}}$$

$$\boldsymbol{h_{\text{tube inside coefficient}} = 205.11\ \text{Btu/ft}^2\ \text{h F}}$$

↓

$$R_{\text{Inside gas film}} = \frac{1}{h_{\text{tube inside coefficient}}}$$

$$R_{\text{Inside gas film}} = \frac{1}{205.11}$$

$$\boldsymbol{R_{\text{Inside gas film}} = 0.0048\ \text{ft}^2\ \text{h F/Btu}}$$

15.4 Metal resistance

$$R_m = \frac{d_{\text{tube}}}{24K_m} \times \ln \frac{d_{\text{tube}}}{(d_{\text{tube}} - t_{\text{tube}})}$$

↓

$$T_{\text{inlet flue gas}} = 2{,}941.93\,^\circ\text{F}$$

15.4.1 Thermal conductivity of metals

$$K_m = 21\,\text{Btu/ft h F}$$

↓

$$R_m = \frac{2}{24 \times 21} \times \ln \frac{2}{(2-0.105)}$$

$$R_m = 0.0002\,\text{ft}^2\,\text{h F/Btu}$$

15.5 Outside gas film resistance

$$R_o = \frac{1}{h_o}$$

$$R_o = \frac{1}{13.36}$$

$$\mathbf{R_o = 0.074\ ft^2\ h\ F/Btu}$$

15.6 Heat flux

$$Q_{flux} = \frac{(T_g - T_{sat.})}{(R_{Inside\ gas\ film} + R_m + R_o)}$$

$$Q_{flux} = \frac{(2,941.93 - 493.8)}{(0.0048 + 0.0002 + 0.074)}$$

$$\mathbf{Q_{flux} = 30,611.36\ Btu/ft^2\ h\ F}$$

15.7 Temperature drop across the gas film

$$\Delta T_{across\ the\ gas\ film} = Q_{flux} \times R_{Inside\ gas\ film}$$

$$\Delta T_{across\ the\ gas\ film} = 30,611.36 \times 0.0048$$

$$\mathbf{\Delta T_{across\ the\ gas\ film} = 2,292.2\ ^\circ F}$$

15.8 Temperature drop across the tube metal

$$\Delta T_{across\ the\ tube\ metal} = Q_{flux} \times R_m$$

$$\Delta T_{across\ the\ tube\ metal} = 30,611.36 \times 0.0002$$

$$\mathbf{\Delta T_{across\ the\ tube\ metal} = 6.55\ ^\circ F}$$

15.9 Temperature drop across the steam film

$$\Delta T_{across\ the\ steam\ film} = Q_{flux} \times R_o$$

$$\Delta T_{across\ the\ steam\ film} = 30,611.36 \times 0.074$$

$$\mathbf{\Delta T_{across\ the\ steam\ film} = 149.23\ ^\circ F}$$

15.10 Average tube metal temperature

$$T_{\text{average tube metal}} = \frac{(T_{\text{g}} - \Delta T_{\text{across the gas film}}) + (T_{\text{sat.}} + \Delta T_{\text{across the steam film}})}{2}$$

$$T_{\text{average tube metal}} = \frac{(2,899.63 - 2,292.2) + (493.8 + 149.23)}{2}$$

$$\boldsymbol{T_{\text{average tube metal}} = 646.38 \ ^\circ\text{F}}$$

16 Furnace draft pressure drop

Draft is named as a difference in pressure between atmospheric pressure and the existing pressure in the furnace or boiler's flue gas passage. Draft can moreover be referred to as pressure difference in chamber area which comes about within the movement of the flue gases and the air stream.

Drafts are created by the rising combustion gases in the stack. For illustration, a blower can be put into four categories: natural, induced, balanced and forced.

- Natural draft: The difference in density of the hot flue gases and cooler surrounding gases creates a pressure differential that moves the hotter flue gases into the cooler surroundings.
- Forced draft: When air or flue gases are kept up over the atmospheric pressure, ordinarily, it is done with the assistance of a forced draft fan.
- Induced draft: When air or flue gases pressure beneath the impact of a continuously decreasing underneath the atmospheric pressure, the system works under induced draft. The stacks give adequate natural draft to meet the low draft loss needs. In an arrangement to meet higher pressure differentials, the stacks must, at the same time, work with draft fans.
- Balanced draft: When atmospheric pressure and static pressure is equal, draft is zero in this system.

Flue gas combustion's rate and the boiler heat transfer amount are both dependent on the movement of flue gases. The rate of combustion increases by strong current of air (draft) through the fuel bed. The heat transfer rate from the flue gases to the boiler increases by stronger movement (which improves efficiency and circulation) [15].

Furnace draft pressure drop is calculated by the following sequence:

- Flue gas density
- Furnace volumetric flow rate
- Furnace flue gas velocity
- Head loss due to dimension change in furnace exit
- Furnace equivalent diameter
- Furnace equivalent length
- Reynolds number in furnace
- Friction factor in furnace
- Furnace draft pressure drop

https://doi.org/10.1515/9783110757088-016

16.1 Flue gas density

$$\rho_g = 492 \times \mathrm{MW} \times \frac{P_{\text{furnace}}}{359 \times (460 + T_{\text{average}}) \times 14.7}$$

$$\rho_g = 492 \times 29 \times \frac{15.53}{359 \times (460 + 2{,}666.23) \times 14.7}$$

$$\boldsymbol{\rho_g = 0.0134\ \mathrm{lb/cu.ft}}$$

16.2 Furnace volumetric flow rate

$$V'_{\text{air}} = \frac{170{,}818.92}{0.0134}$$

$$\boldsymbol{V'_{\text{air}} = 12{,}721{,}318\ (\mathrm{cu.ft/h})}$$

16.3 Furnace flue gas velocity

$$v_{\text{flue gas inside furnace}} = \frac{V'_{\text{air}}/60}{(H \times W)/144}$$

$$v_{\text{flue gas inside furnace}} = \frac{12{,}721{,}318/60}{(100 \times 144)/144}$$

$$\boldsymbol{v_{\text{flue gas inside furnace}} = 2{,}117.2\ \mathrm{fpm}}$$

16.4 Head loss due to dimension change in furnace exit

To exit from furnace to lance area according to Fig. 16.1, $H/W1$ and $WO/W1$ will be 1.442 and 0.4, respectively; then the return correction factor will be 0.53.

SR3-1 Elbow, 90 Degree, Variable Inlet/Outlet Areas, Supply Air Systems

	C_o Values						
	W_o/W_1						
H/W_1	0.6	0.8	1.0	1.2	1.4	1.6	2.0
0.25	0.63	0.92	1.24	1.64	2.14	2.71	4.24
1.00	0.61	0.87	1.15	1.47	1.86	2.30	3.36
4.00	0.53	0.70	0.90	1.17	1.49	1.84	2.64
100.	0.54	0.67	0.79	0.99	1.23	1.54	2.20

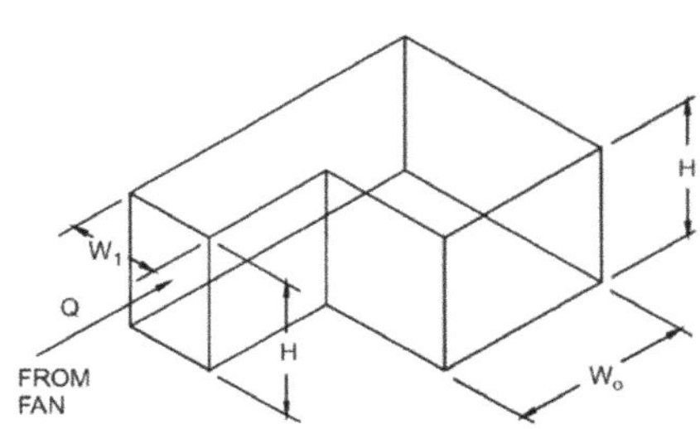

Fig. 16.1: Duct correction factor [16].

To enter from lance area to superheater area according to Fig. 16.1, $H/W1$ and $WO/W1$ will be 1.442 and 0.7, respectively; then contraction correction factor will be 0.53.

So, head loss due to change duct size will be calculated as follows:

$$H_{\text{Loss furnace exit}} = (f_{\text{return}} + f_{\text{contraction}}) \times \frac{(v_{\text{flue gas inside furnace}}/60)^2}{2g}$$

$$H_{\text{Loss furnace exit}} = (0.53 + 0.53) \times \frac{(2{,}1117.2/60)^2}{2 \times 32.2}$$

$$\boldsymbol{H_{\text{Loss furnace exit}} = 20.496\ \text{ft}}$$

16.5 Furnace equivalent diameter

When engineers want to find the proper duct size, they should first find the equivalent diameter which is the diameter of a circular duct with the same pressure loss as an equivalent rectangular duct [5]:

$$d_i = \frac{2WH}{W + H}$$

$$d_i = \frac{2 \times 100 \times 144}{100 + 144}$$

$$\boldsymbol{d_i = 118\ \text{in}}$$

16.6 Furnace equivalent length

Who is looking to ensure proper airflow distribution with the “equal friction method” of duct sizing such as duct system designer try to find equivalent lengths. The equivalent lengths of all fittings and measured distances of straight duct should be calculated [17]:

$$L_e = L_{\text{furnace}} + H_{\text{Loss furnace exit}}$$

$$L_e = \frac{440}{12} + 20.496$$

$$\boldsymbol{L_e = 57\ \text{ft}}$$

16.7 Reynolds number in furnace

$$\text{Re} = 15.2 \times \frac{W'}{\mu \times d_i}$$

$$\mathrm{Re} = 15.2 \times \frac{170,818.92}{0.136 \times 118}$$

$$\mathbf{Re = 161{,}655.71}$$

16.8 Friction factor in furnace

$$\mathrm{fr} = \frac{0.316}{\mathrm{Re}^{0.25}}$$

$$\mathrm{fr} = \frac{0.316}{161{,}655.71^{0.25}}$$

$$\mathbf{fr = 0.016}$$

16.9 Furnace draft pressure drop

$$\Delta P_{\mathrm{furnace}} = 93 \times 10^{-6} \times \mathrm{fr} \times W'^{2} \times \frac{L_{\mathrm{e}}}{\rho d_i^{5}}$$

$$\Delta P_{\mathrm{furnace}} = 93 \times 10^{-6} \times 0.016 \times 170,818.92^{2} \times \frac{57}{0.0134 \times 118^{5}}$$

$$\mathbf{\Delta P_{furnace} = 0.0079\ in\ WG = 0.201\ MM\ WG}$$

17 Boiler design pressure

Design pressure is related to matching temperature and pressure that is expected in a normal condition. Maximum allowable working pressure is a top pressure in normal operating conditions that are allowed in equipment or a vessel [18].

Boiler design pressure will be calculated by the following sequence:
- Steam drum saturated pressure and temperature
- Steam drum first safety valve setting pressure
- Steam drum second safety valve setting pressure
- Boiler design pressure

17.1 Steam drum saturated pressure and temperature

Boiler operating pressure is known as the working pressure and refers to steam drum operating pressure for water tube boilers and shell operating pressure for fire tube boilers.

Steam drum pressure can be calculated by the following assumption:

Superheater outlet pressure	42.82	kg/cm^2
Pressure drops in main stem stop value + nonreturn valve	0.92	kg/cm^2
Pressure drops in main header	0.009	kg/cm^2
Spray attemperator loss	0.34	kg/cm^2
Pressure drops in superheater bank	0.12	kg/cm^2
Pressure drops in boiler bank	0.001	kg/cm^3
Steam drum operating pressure	44.22	kg/cm^2

Please note that all the above mentioned pressure drop is calculated and can be used but for first iteration, we can estimate and calculate the drum pressure and finally by calculating exact data going to find drum pressure. Steam drum pressure is saturated, so drum and bank tube water temperature is equal to 493.8 °F.

17.2 Steam drum first safety valve setting pressure

$$P_{\mathrm{PSV}_1} = P_{\mathrm{drum}} \times 1.05 \times 1.03$$

$$P_{\mathrm{PSV}_1} = 44.22 \times 1.05 \times 1.03$$

https://doi.org/10.1515/9783110757088-017

$$\mathbf{P_{PSV_1} = 47.82\ kg/cm^2}$$

17.3 Steam drum second safety valve setting pressure

$$P_{\mathrm{PSV}_2} = P_{\mathrm{PSV}_1} + 0.34$$

$$P_{\mathrm{PSV}_2} = 47.82 + 0.34$$

$$\mathbf{P_{PSV_2} = 48.16\ kg/cm^2}$$

17.4 Boiler design pressure

$$P_{\mathrm{Design}} = P_{\mathrm{PSV}_2}$$

$$\mathbf{P_{Design} = 48.16\ \ kg/cm^2}$$

18 Superheater package

Superheaters increase the temperature of saturated steam to supply the required process condition. Superheaters are single-phase heat exchangers in which steam flows inside the tubes and flue gas flows outside the tube.

In the steam drum area at boiling point, the saturated steam is separated from water and goes through the superheater tubes. Superheaters' surface temperatures are higher than the boiler tubes. The material of tubes depends on the desired steam temperature. Carbon steel tube will be applicable till 400 °C and chrome-moly steel till 660 °C. Designing is based on the type of heat transfer, convection, radiation or a combination of the two and the superheating circulation. Counterflow, parallel flow and combined parallel and counterflow can be used. Figure 18.1 shows superheater circuits with different flow patterns [19].

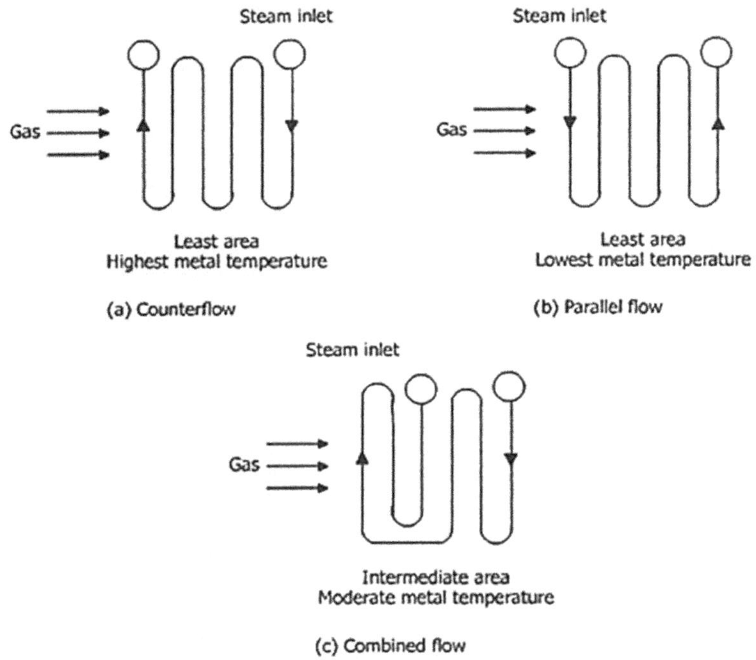

Fig. 18.1: Superheater circuit pattern.

https://doi.org/10.1515/9783110757088-018

A radiant superheater will be used in radiant area of the furnace to absorb heat by radiation. A convection superheater is placed in convection area of combustion chamber which is ordinarily higher than the economizer (Fig. 18.2).

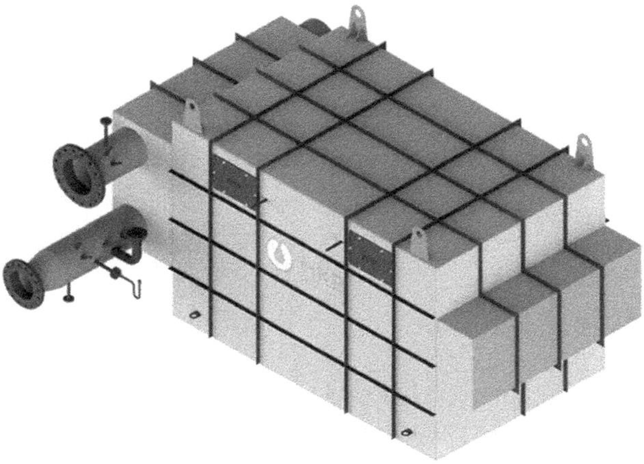

Fig. 18.2: Superheater by HKB Boiler Solution.

19 Superheater tube rows and deep number

To determine the superheater package's tube row and deep number, you will use the following formula. This formula needs selected steam velocity which can be obtained from the velocity table. Superheater tube rows and deep number will be calculated by the following sequence:

- Superheater heat duty prediction
- Superheater tube thickness
- Superheater tube area
- Superheater tube rows and deep number

19.1 Superheater heat duty prediction

A portion of heat has to be transferred from the hot side to the cold side over a unit of time expressed as heat. First, we should calculate the superheater package heat duty as per the following formulas:

Superheater steam inlet temperature	493.8	°F
Superheater inlet pressure	629.01	psig
Superheater inlet enthalpy	1,202.95	Btu/lb
Superheater steam outlet temperature (controlled)	788	°F
Superheater outlet pressure	609.15	psig
Superheater outlet enthalpy	1,400.14	Btu/lb
Steam mass flow	132,277.2	lb/h

$$Q_{duty} = (h_{outlet\ steam} - h_{inlet\ steam}) \times \text{evaporation @ 100\% MCR}$$

$$Q_{duty} = (1{,}400.14 - 1{,}202.95)\ \text{Btu/lb} \times 132{,}277.2\ \text{lb/h}$$

$$\mathbf{Q_{duty} = 26.08\ MMBtu/h}$$

19.2 Superheater tube thickness

To determine the minimum required thickness of tubing, you will use the following formula [20]:

$$t_{min} = \frac{PD}{2S + P} + 0.005\,D + e$$

https://doi.org/10.1515/9783110757088-019

where t is the minimum required tube thickness, in; P is the boiler design pressure, psig; D is the tube diameter, in; S is the maximum allowable stress for A213-T11, psi; e is the thickness factor for expanded tube ends as per ASME 1, PG-27.4

$$t_{\min} = \frac{48.17 \times 14.2233 \times 1.5}{2 \times 6{,}850 + 48.17 \times 14.2233} + 0.005 \times 1.5 + 0.04$$

$$\mathbf{t_{min} = 0.119\ in}$$

As per tube manufacturer catalog for 1.5 in diameter, they produce tube with thickness of 0.12 in. So:

$$\mathbf{t_{selected} = 0.12\ in}$$

19.3 Superheater tube area

$$A_{\text{superheater tube}} = \frac{\pi}{144} \times \left(\frac{D-t}{2}\right)^2$$

$$A_{\text{superheater tube}} = \frac{\pi}{144} \times \left(\frac{1.5-0.12}{2}\right)^2$$

$$A_{\text{superheater tube}} = 0.0103\,\text{ft}^2$$

19.4 Superheater tube rows and deep number

Steam velocity range, 2,000–5,000 fpm [41]

Selected steam velocity = 2,000 fpm

Saturated steam specific volume = 0.709 ft^3/lb

Superheater steam volumetric flow rate = 93,790.3 ft^3/h

Selected tube row deep no. = 4

$$\text{Tube row no.} = \text{round}_{\text{up}}\left(\frac{V'_{\text{superheater steam}}}{V_{\text{selected}} \times 60 \times (\text{tube row no.})}\right)$$

$$\text{Tube row no.} = \text{round}_{\text{up}}\left(\frac{93{,}790.3}{2{,}000 \times 60 \times 4}\right)$$

$$\textbf{Tube row no.} = \mathbf{19}$$

20 Superheater convective heat transfer coefficient prediction

A proportionality constant between the heat flux and the thermodynamic driving force for the flow of heat is called as heat transfer coefficient [21].

Superheater convective heat transfer coefficient will be calculated by the following sequence:

- Superheater flue gas outlet temperature prediction
- Log mean temperature difference (LMTD) prediction
- Superheater average flue gas temperature prediction
- Superheater average flue gas properties
- Superheater tube longitudinal and transverse pitch
- Superheater package long
- Superheater primary heat surface area
- Superheater gas mass velocity
- Superheater convective heat transfer coefficient

20.1 Superheater flue gas outlet temperature prediction

$$Tg_{\text{outlet}} = Tg_{\text{inlet}} - \frac{Q_{\text{superheater}}}{m'_{\text{flue gas}} \times Cp_{\text{average gas temperature}} \times (1 - \text{heat loss}/100)}$$

At this step, we need to find the primary flue gas outlet temperature and specific heat at average temperature, and then by repeating this sequence, flue gas outlet temperature can be found [5]:

$$Tg_{\text{outlet}} = 2{,}390.5 - \frac{26.08 \times 10^6}{170{,}818.92 \times 0.3449 \times (1 - 2/100)}$$

$$\boldsymbol{Tg_{\text{outlet}} = 1{,}938.76^\circ\text{F}}$$

20.2 Log mean temperature difference prediction

In heat transfer area such as heat exchangers, temperature driving force can be determined as LMTD (also known as log mean temperature difference). For a fixed size and heat transfer coefficient, heat exchangers transferred more heat by larger LMTD [22]:

https://doi.org/10.1515/9783110757088-020

$$\Delta T_{\log} = \frac{\left(Tg_{\text{outlet}} - T_{\text{outlet steam}}\right) - \left(Tg_{\text{inlet}} - T_{\text{Inlet steam}}\right)}{\text{LN}\frac{Tg_{\text{outlet}} - T_{\text{outlet steam}}}{Tg_{\text{inlet}} - T_{\text{Inlet steam}}}}$$

$$\Delta T_{\log} = \frac{(1,938.76 - 788) - (2,390.5 - 493.8)}{\text{LN}\frac{1,938.76 - 788}{2,390.5 - 493.8}}$$

$$\boldsymbol{\Delta T_{\log} = 1.49^{\circ}\text{F}}$$

20.3 Superheater average flue gas temperature prediction

$$T_{\text{average}} = \frac{Tg_{\text{inlet}} + Tg_{-\text{Predicted outlet}}}{2}$$

$$T_{\text{average}} = \frac{1,938.76 + 2,390.5}{2}$$

$$\boldsymbol{T_{\text{average}} = 2,164.64^{\circ}\text{F}}$$

20.4 Superheater average flue gas properties

Superheater average flue gas properties can be found from gas tables at average temperature:

$$\mu = 0.1298\ \text{lb/ft h}$$

$$C_p = 0.3449\ \text{Btu/lb F}$$

$$K = 0.0585\ \text{Btu/ft}\,h\,F$$

20.5 Superheater tube longitudinal and transverse pitch

In a convection surface except heat absorption and resistance to gas flow, other factors such as optimum tube spacing and arrangement must be considered.

For designing superheater package, first, transverse and longitudinal space will be selected, and by checking performance result and draft pressure loss will find optimum space that required.

At this step, we consider transverse and longitudinal pitches as 4 and 4.03125, respectively.

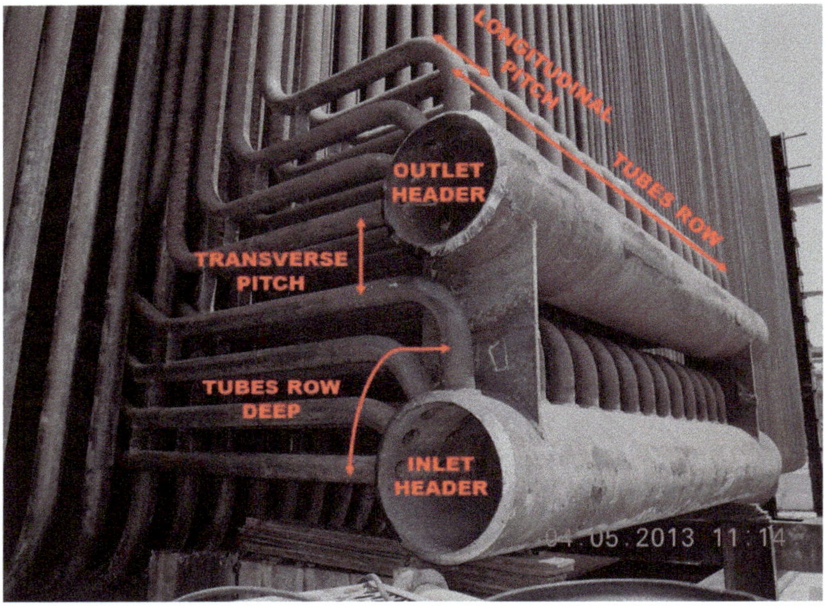

Fig. 20.1: Water tube boiler (132 T/h, 42 barg, 390 °C), Basra Petrochemical, Basra, Iraq.

20.6 Superheater package long

$$L_{\text{superheater}} = \text{Round}[\text{ref. no. } S_{\text{L}} \times (N_{\text{tube row}} - c1)]$$

$$L_{\text{superheater}} = \text{Round}[\text{ref. no. } 4.03125 \times (19 - 1)]$$

$$\mathbf{L_{\text{super heater}} = 73 \quad in}$$

20.7 Superheater primary heat surface area

Tube length related to superheater duty and outlet temperature need to be optimum by checking primary length selection. By assuming each tube length as 40 ft, heat surface is given by

$$A_{\text{superheater}} = \pi \times D_{\text{tube}}/12 \times N_{\text{tube row}} \times N_{\text{tube row deep}} \times L_{\text{superheater tube}}$$

$$A_{\text{superheater}} = \pi \times 1.5/12 \times 4 \times 19 \times 40$$

$$\mathbf{A_{\text{superheater}} = 1,193.2\,ft^2}$$

20.8 Superheater gas mass velocity

Gas mass velocity is mass flow rate across a unit area vertical to the direction of the velocity vector and can be find as follows [5]:

$$G = 12\frac{W'}{N_{\text{row deep}} \times L_{\text{tube}} \times (S_{\text{T}} - d)}$$

$$G = 12 \times \frac{170,818.92}{4 \times 40 \times (4 - 1.5)}$$

$$\mathbf{G = 5,124.57\ lb/ft^2\ h}$$

20.9 Superheater convective heat transfer coefficient

$$h_{\text{C}} = 0.9 \times \frac{G^{0.6}}{D_{\text{tube}}{}^{0.4}} \times \frac{K^{0.67} \times C_P{}^{0.33}}{\mu^{0.27}}$$

$$h_{\text{C}} = 0.9 \times \frac{5,124.57^{0.6}}{1.5^{0.4}} \times \frac{0.0585^{0.67} \times 0.3349^{0.33}}{0.1298^{0.27}}$$

$$\mathbf{h_{C} = 23.47\ Btu/ft^2\ h\ F}$$

21 Superheater uncontrolled outlet steam temperature prediction

All generated steam at boilers must go through the superheater. As a result, all D-type boilers are called as uncontrolled superheat boilers. The design characteristics guarantee that the temperature will reach to the set point. The degree of superheat will be defined as differences between steam drum temperature and the actual reading on the superheater outlet temperature gage [23].

Superheater uncontrolled outlet steam temperature will be calculated by the following sequence:
- Superheater package performance prediction
- Superheater flue gas outlet temperature
- Superheater flue gas average temperature
- Superheater outlet steam temperature
- Each tube steam flow rate
- Inside gas film resistance
- Metal resistance
- Outside gas film resistance
- Heat flux
- Temperature drop across the gas film
- Temperature drop across the tube metal
- Temperature drop across the steam film
- Superheater uncontrolled outlet steam temperature prediction

21.1 Superheater package performance prediction

Prediction of the exit temperatures and duty can be done by number of transfer units (NTU) method. Basically, the duty Q is given by [5]

$$Q_{\text{superheater}} = \varepsilon C_{\min}\left(T_{\text{gas}_{\text{inlet}}} - T_{\text{steam}_{\text{inlet}}}\right)$$

where ε depends on the type of flow, whether counter flow, parallel flow or cross-flow. In superheater, usually a counter-flow arrangement is selected. ε for this is given by

$$\varepsilon = \frac{1 - \exp[-NTU \times (1 - C)]}{1 - C \times \exp[-NTU \times (1 - C)]}$$

where

https://doi.org/10.1515/9783110757088-021

$$NTU = \frac{UA}{C_{min}}$$

$$C = \frac{C_{min}}{C_{max}}$$

$$C_{min} = \left(W'C_p\right)_{min}$$

$$C_{max} = \left(W'C_p\right)_{max}$$

At our design case,

Average steam heat capacity @ constant volume = 0.4373 Btu/lb F

Inlet flue gas heat capacity @ constant volume = 0.3449 Btu/lb F

$W'_{steam} = 132,277.2$ lb/h

$W'_{flue\ gas} = 170,818.92$ lb/h

$$C_{max} = 0.4373 \times 132,277.2 = 57,844.81$$

$$C_{min} = 0.3449 \times 170,818.92 \times (1 - 2/100) = 57,737.13$$

↓

$$C = \frac{57,737.13}{57,844.81} = 0.998$$

Because gas temperature is low, then non-luminous heat transfer is low and can be neglected.

↓

$$\text{NTU} = \frac{23.47 \times 1,193.2}{57,737.13} = 0.48$$

↓

$$\varepsilon = \frac{1 - \exp[-0.48 \times (1 - 0.998)]}{1 - 0.998 \times \exp[-0.48 \times (1 - 0.998)]} = 0.326$$

↓

$$Q_{\text{superheater}} = 0.326 \times 57,737.13 \times (2390.5 - 494.8)$$

$$\boldsymbol{Q_{\text{superheater}} = 35.77\ \text{MM Btu/h}}$$

21.2 Superheater flue gas outlet temperature

$$T_{\text{gas}_{\text{outlet}}} = T_{\text{gas}_{\text{inlet}}} - \frac{Q}{C_p \times W'_{\text{gas}} \times (1 - \text{heat loss\%})}$$

$$T_{\text{gas}_{\text{outlet}}} = 2,390.5 - \frac{35.77 \times 10^6}{0.3449 \times 170,818.92 \times (1 - \frac{2}{100})}$$

$$\boldsymbol{T_{\text{gas}_{\text{outlet}}} = 1,750.96\ ^{\circ}\text{F}}$$

21.3 Superheater flue gas average temperature

$$T_{\text{average flue gas}} = \frac{T_{\text{outlet flue gas}} + T_{\text{flue gas inlet}}}{2}$$

$$T_{\text{average flue gas}} = \frac{1,750.96 + 2,390.5}{2}$$

$$\boldsymbol{T_{\text{average flue gas}} = 2,070.74\ ^{\circ}\text{F}}$$

21.4 Superheater outlet steam temperature

$$T_{\text{steam}_{\text{outlet}}} = T_{\text{steam}_{\text{inlet}}} + \frac{Q}{C_p \times W'_{\text{steam}}}$$

$$T_{\text{steam}_{\text{outlet}}} = 494.8 + \frac{35.77 \times 10^6}{0.4376 \times 132,277.2}$$

$$\boldsymbol{T_{\text{steam}_{\text{outlet}}} = 1,112.31\ ^{\circ}\text{F}}$$

So, outlet temperature is bigger and near to what we expect from our superheater design which is acceptable. If not, it should change the tube diameter or tube length, … .

21.5 Each tube steam flow rate

First, we should find each tube steam mass flow which is equal to the output steam mass flow rate divided by the superheater total tube numbers:

$$m'_{\text{each tube}} = \frac{m'_{\text{output steam}}}{N_{\text{boiler tube}}}$$

$$m'_{\text{each tube}} = \frac{132{,}277.2}{4 \times 19}$$

$$m'_{\text{each tube}} = 1{,}740.49 \ \text{lb/h}$$

21.6 inside gas film resistance

$$R_{\text{Inside gas film}} = \frac{1}{h_{\text{tube inside coefficient}}}$$

↓

$$h_{\text{tube inside coefficient}} = \frac{2.44 \times m'^{\,0.8}_{\text{each tube}} \times C}{(d_{\text{tube}} - t_{\text{tube}})^{1.8}}$$

↓

C factor will be obtained from Tab. 21.1 by drum saturated pressure and temperature, in which drum saturated pressure is superheater output pressure plus superheater and boiler bank and another pressure losses. Here, total pressure loss can be considered as 1–6 barg and after final designing it can be optimized by iteration:

$$P_{\text{average superheater}} = \frac{P_{\text{superheater outlet}} + P_{\text{superheater inlet}}}{2} + 14.7$$

$$P_{\text{average superheater}} = \frac{609.15 + 629.01}{2} + 14.7$$

$$\boldsymbol{P_{\text{average superheater}} = 633.78 \ \text{psia}}$$

↓

$$T_{\text{superheater outlet}} = 1,112.31\ ^{\circ}\text{F}$$

↓

$$C = 0.347$$

Tab. 21.1: Factor *C* for steam [5].

Temperature (°F)	Drum saturation pressure (psia)				
	100	200	500	1,000	2,000
400	0.2716	0.3059			
500	0.2737	0.2909	0.3595		
600	0.2813	0.2896	0.3228	0.413	
700	0.2917	0.2965	0.3161	0.3586	0.5206
800	0.3050	0.3090	0.3206	0.3453	0.4214
900	0.3161	0.3197	0.3277	0.3477	0.3946
1,000	0.3276	0.3302	0.3392	0.3531	0.386

↓

$$h_{\text{tube inside coefficient}} = \frac{2.44 \times 1,740.49^{0.8} \times 0.347}{(1.5 - 0.12)^{1.8}}$$

$$h_{\text{tube inside coefficient}} = 185.55\ \text{Btu/ft}^2\ \text{h F}$$

↓

$$R_{\text{Inside gas film}} = \frac{1}{h_{\text{tube inside coefficient}}}$$

$$R_{\text{Inside gas film}} = \frac{1}{185.55}$$

$$R_{\text{Inside gas film}} = 0.005\ \text{ft}^2\ \text{h F/Btu}$$

21.7 Metal resistance

$$R_{\mathrm{m}} = \frac{d_{\mathrm{tube}}}{24K_{\mathrm{m}}} \times \ln \frac{d_{\mathrm{tube}}}{(d_{\mathrm{tube}} - t_{\mathrm{tube}})}$$

↓

$$\boldsymbol{T_{\text{average flue gas}} = 2,070.74\ {}^{\circ}\mathrm{F}}$$

21.7.1 Thermal conductivity of metals

$$K_{\mathrm{m}} = 15\ \mathrm{Btu/ft\,h\,F}$$

↓

$$R_{\mathrm{m}} = \frac{2}{24 \times 15} \times \ln \frac{2}{(2 - 0.105)}$$

$$\boldsymbol{R_{\mathrm{m}} = 0.0003\ \mathrm{ft^2\,h\,F/Btu}}$$

21.8 Outside gas film resistance

$$R_{\mathrm{o}} = \frac{1}{h_{\mathrm{o}}}$$

$$R_{\mathrm{o}} = \frac{1}{23.47}$$

$$\boldsymbol{R_{\mathrm{o}} = 0.042\ \mathrm{ft^2\,h\,F/Btu}}$$

21.9 Heat flux

$$Q_{\mathrm{flux}} = \frac{(T_{\text{average flue gas}} - T_{\text{steam outlet}})}{(R_{\text{Inside gas film}} + R_{\mathrm{m}} + R_{\mathrm{o}})}$$

$$Q_{\mathrm{flux}} = \frac{(2,070.74\ - 1,112.31)}{(0.005 + 0.0003 + 0.042)}$$

$$\mathbf{Q_{flux} = 19,822.42\,Btu/ft^2\,h\,F}$$

21.10 Temperature drop across the gas film

$$\Delta T_{\text{across the gas film}} = Q_{\text{flux}} \times R_{\text{Inside gas film}}$$

$$\Delta T_{\text{across the gas film}} = 19,822.42 \times 0.005$$

$$\boldsymbol{\Delta T_{\textbf{across the gas film}} = 844.7\ ^\circ\mathbf{F}}$$

21.11 Temperature drop across the tube metal

$$\Delta T_{\text{across the tube metal}} = Q_{\text{flux}} \times R_{\text{m}}$$

$$\Delta T_{\text{across the tube metal}} = 19,822.42 \times 0.0003$$

$$\boldsymbol{\Delta T_{\textbf{across the tube metal}} = 6.88\ ^\circ\mathbf{F}}$$

21.12 Temperature Drop across the steam film

$$t\Delta T_{\text{across the steam film}} = Q_{\text{flux}} \times R_{\text{o}}$$

$$\Delta T_{\text{across the steam film}} = 19,822.42 \times 0.042$$

$$\boldsymbol{\Delta T_{\textbf{across the steam film}} = 106.827\ ^\circ\mathbf{F}}$$

21.13 Superheater uncontrolled outlet steam temperature prediction

$$T_{\text{outlet steam}} = T_{\text{steam}_{\text{outlet}}} - \Delta T_{\text{across the gas film}} - \Delta T_{\text{across the tube metal}} - \Delta T_{\text{across the steam film}}$$

$$T_{\text{outlet steam}} = 1,112.31 - 844.7 - 6.88\ - 106.827$$

$$\boldsymbol{T_{\textbf{outlet steam}} = 792.54\ ^\circ\mathbf{F} = 422.52\ \mathbf{C}}$$

22 Superheater flue gas draft pressure drop

Superheater flue gas draft pressure drop will be calculated by the following sequence:
- Superheater flue gas density
- Superheater flue gas draft Reynolds number
- Superheater flue gas draft friction factor for inline arrangement
- Superheater flue gas draft pressure drop

22.1 Superheater flue gas density

$$\rho_{\text{flue gas}} = 492 \times 29.24 \times \frac{15.53}{359 \times (460 + 2{,}072.26 \times 14.7)}$$

$$\boldsymbol{\rho_{\text{flue gas}} = 0.0167\ \text{lb/cu.ft}}$$

22.2 Superheater flue gas draft Reynolds number

$$\text{Re} = \frac{Gd_{\text{Tube}}}{12\ \mu}$$

$$\text{Re} = \frac{5{,}124.57 \times 1.5}{12 \times 0.12}$$

$$\mathbf{Re = 5{,}385.21}$$

23.3 Superheater flue gas draft friction factor for inline arrangement

For calculating the flue gas friction loss, use a dimensionless value known as the friction factor and can be found by the following formula [5]:.

$$\text{fr} = \text{Re}^{-0.15} \times \left(0.044 + \frac{0.08 \times \text{S}_\text{L}/\text{d}_\text{tube}}{(\text{S}_\text{T}/\text{d}_\text{tube} - 1)^{0.43 + 1.13 \times \text{d}_\text{tube}/\text{S}_\text{L}}}\right)$$

$$\text{fr} = 5{,}385.21^{-0.15} \times \left(0.044 + \frac{0.08 \times 4.03125/1.5}{(4/1.5 - 1)^{0.43 + 1.13 \times 1.5/4.03125}}\right)$$

$$\mathbf{fr = 0.051}$$

https://doi.org/10.1515/9783110757088-022

22.4 Superheater flue gas draft pressure drop

$$\Delta P_g = 9.3 \times 10^{-10} \times G^2 \times \text{fr} \times \frac{N_{\text{tube row}}}{\rho}$$

$$\Delta P_g = 9.3 \times 10^{-10} \times 5{,}385.21^2 \times 0.051 \times \frac{19}{0.0167}$$

$$\Delta P_g = 1.40 \text{ in WG} = 35.60 \text{ mm WG}$$

23 Superheater package total steam pressure drop

Superheater package total steam pressure drop will be calculated by the following sequence:

- Superheater inlet header diameter
- Superheater inlet header design pressure
- Superheater inlet header thickness by pressure
- Superheater inlet header thickness by tube holes
- Superheater inlet header selected thickness
- Superheater outlet header diameter and thickness
- Superheater inlet header Reynolds number
- Superheater inlet header friction factor
- Superheater inlet header steam pressure drop
- Superheater outlet header Reynolds number
- Superheater outlet header friction factor
- Superheater outlet header steam pressure drop
- Superheater tube bundle Reynolds number
- Superheater tube bundle friction factor
- Superheater tube bundle steam pressure drop
- Superheater package total steam pressure drop

23.1 Superheater inlet header diameter

The superheater header's diameter can be determined by assuming steam velocity and then controlling the steam velocity.

High-pressure steam velocity range inside boiler, 40.6–61 m/s [41]
Selected steam velocity = 61 m/s
Saturated steam flow = 60/3.6 = 16.66 kg/s
Saturated steam-specific volume = 0.044 m^3/kg
Superheater steam volumetric flow rate = 0.737 m^3/s
Area required = 0.737/61 = 0.012 m^2
Required header diameter = 4.88 in
Selected header diameter = 6 in

$$V = \frac{0.737}{\pi/4 \times \left(\frac{6 \times 25.4}{1,000}\right)^2} = 40.46 \ \text{m/s}$$

Then the diameter selection is okay.

https://doi.org/10.1515/9783110757088-023

23.2 Superheater inlet header design pressure

The design pressure will be defined according to the following criteria except in cases approved by the company [24]:

- For maximum normal operating pressure less than 1.5 barg, use 3.5 bar gage.
- For maximum normal operating pressures between 1.5 and 20 barg, use the maximum normal operating gage pressure +2 bar.
- For maximum normal operating pressures between 20 and 80 barg, use 110% of the maximum normal operating gage pressure.

Drum operating pressure = 44.22 kg/cm^2 = 43.36 barg

Superheater inlet header design pressure: 43.36 × 1.1 = 47.7 barg

23.3 Superheater inlet header thickness by pressure

Minimum required thickness of header can be determined by tow method, and by pressure value, it can be used by the following formula [20]:

$$t_{\min_1} = \frac{PD}{2 \times (68.947 \times S \times E + P \times Y)} + A$$

where t is the minimum required header thickness, in; P is the header design pressure, barg; D is the header diameter, mm; S is the maximum allowable stress for A106-Gr.B, Ksi; E is joint efficiency; Y is temperature coefficient; A is corrosion allowance, mm

$$t_{\min_1} = \frac{47.7 \times 6 \times 25.4}{2 \times (68.947 \times 15 \times 0.85 + 47.7 \times 0.4)} + 3$$

$$t_{\min_1} = 7.04\ \text{mm} = 0.277\ \text{in}$$

23.4 Superheater inlet header thickness by tube holes

Minimum required thickness of header can be determined by another method which is ligament method and it can be used by the following formula [25]:

- ASME SEC.1 – PG-52.4 Diagonal efficiency

$$\text{PG52.4 Diagonal efficiency} = \frac{S_L - d_{\text{tube hole}}}{S_L}$$

$$\text{PG52.4 Diagonal efficiency} = \frac{4.03125 - (1.5 + 0.03125)}{4.03125}$$

PG52.4 Diagonal efficiency = 0.620

$$t_{\min_2} = \frac{P \times D}{2 \times (S \times \text{Eff}_{\text{diagonal}} + Y \times P)} + \text{C.A.}$$

$$t_{\min_2} = \frac{47.7 \times 14.5038 \times 6}{2 \times (15 \times 1,000 \times 0.620 + 0.4 \times 47.7 \times 14.5038)} + 3/25.4$$

$$\mathbf{t_{\min_2} = 0.335\ in}$$

23.5 Superheater inlet header selected thickness

Between two required thicknesses calculated from pressure and ligament method, minimum required thickness will be the bigger one. So $t_{\min_2} > t_{\min_1}$, then minimum required thickness will be 0.337 in, which we can select 0.344 in thickness or schedule 40 for 6 in diameter header from pipe manufacturer catalog.

23.6 Superheater outlet header diameter & thickness

Similar to the above mentioned calculation, outlet header which is drive by superheated steam should be A335-P11 material be 13.6 maximum allowable stress

Superheater outlet header diameter = 8 in
Superheater outlet header thickness = 0.5
Superheater outlet header schedule = 80

23.7 Superheater inlet header Reynolds number

$$\text{Re} = 15.2 \times \frac{W'}{\mu \times d_i}$$

$$\text{Re} = 15.2 \times \frac{132,277.2}{0.0429 \times (6 - 2 \times 0.344)}$$

$$\mathbf{Re = 8,806.52}$$

23.8 Superheater inlet header friction factor

The friction factor of turbulent flow of steam inside header is given by

$$\text{fr} = \frac{0.316}{\text{Re}^{0.25}}$$

$$fr = \frac{0.316}{8,806.52^{0.25}}$$

$$\textbf{fr} = \mathbf{0.006}$$

23.9 Superheater inlet header steam pressure drop

$$\Delta P_\text{g} = 3.36 \times 10^{-6} \times \text{fr} \times L_\text{inlet header} \times \frac{W'^2}{\rho \times (D - t)^5}$$

↓

$$L_\text{inlet header} = \frac{S_\text{L} \times N_\text{tube row}}{12} + 2$$

↓

$$\Delta P_\text{g} = 3.36 \times 10^{-6} \times 0.006 \times \left(\frac{19 \times 4.03125}{12} + 2\right) \times \frac{132,277.2^2}{0.709 \times (6 - 0.344)^5}$$

$$\boldsymbol{\Delta P_\text{g}} = \mathbf{0.35\ psig = 2.41\ kpag}$$

23.10 Superheater outlet header Reynolds number

$$\text{Re} = 15.2 \times \frac{W'}{\mu \times d_i}$$

$$\text{Re} = 15.2 \times \frac{132,277.2}{0.061 \times (8 - 2 \times 0.5)}$$

$$\textbf{Re} = \mathbf{4,704.23}$$

23.11 Superheater outlet header friction factor

The friction factor of turbulent flow of steam inside the header is given by

$$\text{fr} = \frac{0.316}{\text{Re}^{0.25}}$$

$$\text{fr} = \frac{0.316}{4,704.23^{0.25}}$$

$$\mathbf{fr = 0.007}$$

23.12 Superheater outlet header steam pressure drop

$$\Delta P_g = 3.36 \times 10^{-6} \times \text{fr} \times L_e \times \frac{W'^2}{\rho \times (D - t)^5}$$

↓

$$L_{\text{outlet header}} = L_{\text{inlet header}} + 3$$

↓

$$\Delta P_g = 3.36 \times 10^{-6} \times 0.007 \times \left(\frac{19 \times 4.03125}{12} + 2 + 3\right) \times \frac{132,277.2^2}{1.12 \times (8 - 0.5)^5}$$

$$\mathbf{\Delta P_g = 0.216\ psig = 1.48\ kpag}$$

23.13 Superheater tube bundle Reynolds number

$$\text{Re} = 15.2 \times \frac{W'}{\mu \times d_i}$$

$$\text{Re} = 15.2 \times \frac{132,277.2/19 \times 4}{0.0429 \times (1.5 - 2 \times 0.12)}$$

$$\mathbf{Re = 488.515}$$

23.14 Superheater tube bundle friction factor

The friction factor of turbulent flow of steam inside the tube is given by

$$\text{fr} = \frac{0.316}{\text{Re}^{0.25}}$$

$$\text{fr} = \frac{0.316}{488.515^{0.25}}$$

$$\mathbf{fr = 0.012}$$

23.15 Superheater tube bundle steam pressure drop

$$\Delta P_g = 3.36 \times 10^{-6} \times \text{fr} \times L_e \times \frac{W'^2}{\rho \times (D-t)^5}$$

↓

$$L_e = L_{tube} + n \times L_{elbow}$$

$$L_{elbow} = 16 \times \frac{D-2t}{12}$$

$$L_e = 40 + 16 \times 16 \times \frac{1.5 - 2 \times 0.12}{12}$$

$$\mathbf{L_e = 66.88\ ft}$$

↓

$$\Delta P_g = 3.36 \times 10^{-6} \times 0.012 \times 66.88 \times \frac{(132,277.2/19 \times 4)^2}{1.249 \times (1.5 - 0.12)^5}$$

$$\mathbf{\Delta P_g = 1.21\ psig = 8.34\ kpag}$$

23.16 Superheater package total steam pressure drop

$$\Delta P_{superheater} = \Delta P_{inlet\ header} + \Delta P_{outlet\ header} + \Delta P_{tube\ bundle}$$

$$\Delta P_{superheater} = 0.35 + 0.216 + 1.21$$

$$\mathbf{\Delta P_{superheater} = 1.78\ psig = 12.24\ kpag}$$

24 Steam and mud drum sizing

For sizing of the steam drum, the following requirements should be considered:
1. Providing adequate space for dismantling and re-installation for internals.
2. When steam request at superheater outlet pressure (SOP) increases, supply sufficient water to prevent from sudden surges. This makes saturated water to rapidly evaporate.

When drum pressure increases for the same function, drum diameters reduce since specific volume of steam at higher pressure dynamically reduces. Normally, in conventional boilers, each steam drums are identified by inside diameter which ranges from ~1,000 to 2,200 mm (~40 to 87 in). In package boilers, usually steam drum sizes are starting at 914 mm (36 in.). Well-known sizes are 1,067 mm (42 in), 1,220 mm (48 in), 1,370 mm (54 in), 1,524 mm (60 in) and, every so often, 1,676 mm (66 in).

Lower drums identify as mud drums because of large volume of sludge. The lower drum is always smaller than the steam drum because there are no vital internals.

In shop-assembled boilers, package boilers, lower drums are usually made of seamless pipes from size 610 mm (24 in). The lowest diameter is 760 mm (30 in). The other well-known diameter is 914 mm (36 in) and, once in a while, 1,067 mm (42 in) [26].

For our boiler capacity (132,277.2 lb/h), it usually considers 48″ and 30″ for steam and mud drum, respectively.

Fig. 24.1: Elliptical bottom [40].

https://doi.org/10.1515/9783110757088-024

Steam and mud drum sizing will be calculated by the following sequence:

- Steam drum tube holes longitudinal and transverse pitch
- Steam drum thickness by pressure method
- Steam drum diagonal pitch
- Required longitudinal efficiency as per ASME SEC.1 PG-27.2.2
- Actual longitude efficiency as per ASME SEC.1 PG-52.4
- Required circumferential efficiency as per ASME SEC.1 PG-52.3
- Actual circumferential efficiency as per ASME SEC.1 PG-52.3
- Verification of the weakest ligament as per ASME SEC.1 PG-52.3
- Diagonal efficiency as per ASME SEC.1 PG-52.3
- Maximum permissible ligament as per ASME SEC.1 PG-52.3
- Steam drum thickness by ligament
- Steam drum selected thickness
- Steam drum ellipsoidal head
- Mud drum selected thickness
- Mud drum ellipsoidal head

24.1 Steam drum tube holes longitudinal and transverse pitch

In a convection surface, except heat absorption and resistance to gas flow, other factors such as optimum tube spacing and arrangement must be considered [5].

For designing a boiler bank, first transverse and longitudinal space will be selected, and by checking the performance result and draft pressure loss, optimum space that is required will be found.

At this step, we consider transverse and longitudinal pitches as 4.239 and 4.092, respectively.

24.2 Steam drum thickness by pressure method

$$t_{\min_1} = \frac{\mathrm{PD}}{2\times(68.947\times S\times E+P\times Y)} + \mathrm{C.A.}$$

where t is the minimum required drum thickness, inch; P is the boiler design pressure, barg; D is the drum diameter, mm; S is the maximum allowable stress for A516-Gr.70, Ksi;

E is the joint efficiency;

Y is the temperature coefficient;

C.A. is corrosion allowance, mm

$$t_{\min_1} = \frac{48.16 \times 0.980665 \times 48 \times 25.4}{2 \times (68.947 \times 18.8 \times 0.85 + 48.16 \times 0.980665 \times 0.4)} + 3$$

$$\boldsymbol{t_{\min_1} = 28.69\ \text{mm} = 1.129\ \text{in}}$$

24.3 Steam drum diagonal pitch

PG-52.3 diagonal pitch

$$S_{\text{diagonal}} = \sqrt{{S_\text{L}}^2 + \left[\theta_{\text{between holes}} \times \frac{\pi(D+t)}{360}\right]^2}$$

$$S_{\text{diagonal}} = \sqrt{4.03125^2 + \left[8 \times \frac{\pi(48+2)}{360}\right]^2}$$

$$\boldsymbol{S_{\text{diagonal}} = 5.33\ \text{in}}$$

24.4 Required longitudinal efficiency as per ASME SEC.1 PG-27.2.2

At this step, it should assume thickness as 1.625 in and by following sequence check the given value, and if verification is okay, then assumed thickness is also fine:

$$E_{\text{Long}_{\text{required}}} = \frac{P \times R}{S \times t} + (1 - Y) \times \frac{P}{S}$$

$$E_{\text{Long}_{\text{required}}} = \frac{48.16 \times 0.980665 \times 14.5038 \times 48/2}{18.8 \times 1{,}000 \times 2} + (1 - 0.4) \times 43.13 \times 14.5038/(18.8 \times 1000)$$

$$\boldsymbol{E_{\text{Long}_{\text{required}}} = 0.459\%}$$

24.5 Actual longitude efficiency as per ASME SEC.1 PG-52.4

$$E_{\text{Long}_{\text{actual}}} = \frac{S_\text{L} - d_{\text{tube hole}}}{S_\text{L}}$$

$$E_{\text{Long}_{\text{actual}}} = \frac{4.03125 - 2.3125}{4.03125}$$

$$\boldsymbol{E_{\text{Long}_{\text{actual}}} = 0.496\%}$$

Actual longitude efficiency = 0.496 > required longitudinal efficiency = 0.459

Then, longitudinal efficiency meets the design requirement.

24.6 Required circumferential efficiency as per ASME SEC.1 PG-52.3

The required circumferential ligament between tube holes should be minimum one-half the required longitudinal ligament between the tube.

$$E_{\text{circumferential}_{\text{required}}} = \frac{E_{\text{Long}_{\text{required}}}}{2}$$

$$E_{\text{circumferential}_{\text{required}}} = \frac{0.459}{2}$$

$$\boldsymbol{E_{\text{circumferential}_{\text{required}}} = 0.229\%}$$

24.7 Actual circumferential efficiency as per ASME SEC.1 PG-52.3

$$E_{\text{circumferential}_{\text{actual}}} = \frac{S_{\text{T}} - d_{\text{tube hole}}}{S_{\text{T}}}$$

$$E_{\text{circumferential}_{\text{actual}}} = \frac{4.239 - 2.3125}{4.239}$$

$$\boldsymbol{E_{\text{circumferential}_{\text{actual}}} = 0.521\%}$$

Actual circumferential efficiency = 0.521 > required circumferential efficiency = 0.229

Then, longitudinal efficiency meets the design requirement.

24.8 Verification of the weakest ligament as per ASME SEC.1 PG-52.3

$$E_{\text{Long}_{\text{actual}}}/2 = \frac{0.496}{2} = 0.248 < E_{\text{circumferential}_{\text{actual}}} = 0.521$$

Then, the longitudinal efficiency is weakest. So:

$$\boldsymbol{E_{\text{long}} = 0.496\%}$$

24.9 Diagonal efficiency as per ASME SEC.1 PG-52.3

$$E_{\text{diagonal}} = \frac{S_{\text{diagonal}}}{S_{\text{Longitudinal}}}$$

$$E_{\text{diagonal}} = \frac{5.33}{4.03125}$$

$$\boldsymbol{E_{\text{diagonal}} = 1.323\%}$$

24.10 Maximum permissible ligament as per ASME SEC.1 PG-52.3

Figure 24.2 is applicable for openings of definite pattern in pressure parts.

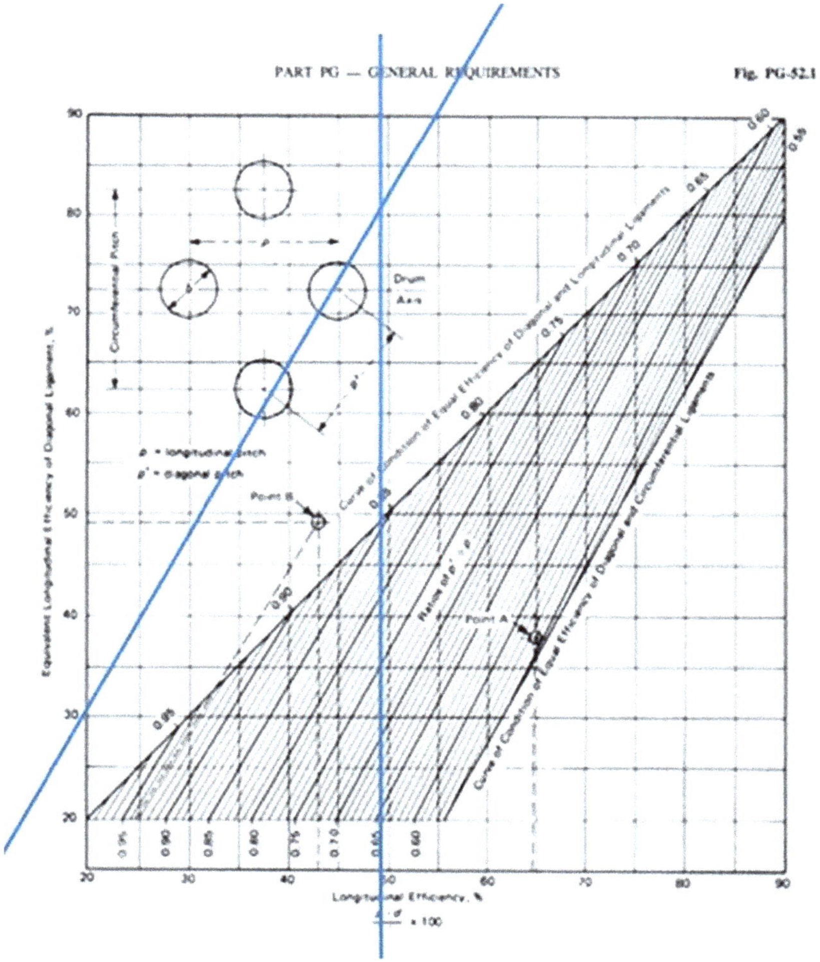

Fig. 24.2: 132 T/h, 42 barg, 390 °C water tube boiler, Basra Petrochemical, Basra, Iraq.

As per curve Fig. 52.1 is equivalent to long efficiency diagonal ligament (%) = 0.8

Then, the maximum permissible ligament will be minimum of longitudinal and diagonal and equivalent diagonal which is 0.496.

24.11 Steam drum thickness by ligament

$$t_{\min_2} = \frac{\text{PD}}{2\times[S\times E-(1-Y)\times P]} + \text{C.A.}$$

$$t_{\min_2} = \frac{48.16\times 0.980665\times 14.5038\times 48}{2\times[18.8\times 1,000\times 0.496-(1-0.4)\times 48.16\times 0.980665\times 14.5038]} + \frac{3}{25.4}$$

$$\boldsymbol{t_{\min_2} = 1.962 \text{ in}}$$

24.12 Steam drum selected thickness

Between the two required thicknesses calculated from pressure and ligament method, minimum required thickness will be the bigger one. So, $t_{\min_2} > t_{\min_1}$; then, the minimum required thickness will be 1.962 in, for which from pipe manufacturer catalog we can select 2 in thickness for 48 in diameter steam drum.

24.13 Steam drum ellipsoidal head

Boiler manufacturers develop elliptical bottoms in two conceivable variations:

a) elliptical bottoms 1.9:1 per NFE 81–103

b) elliptical bottoms 2:1 per ASME VIII Div. 1.

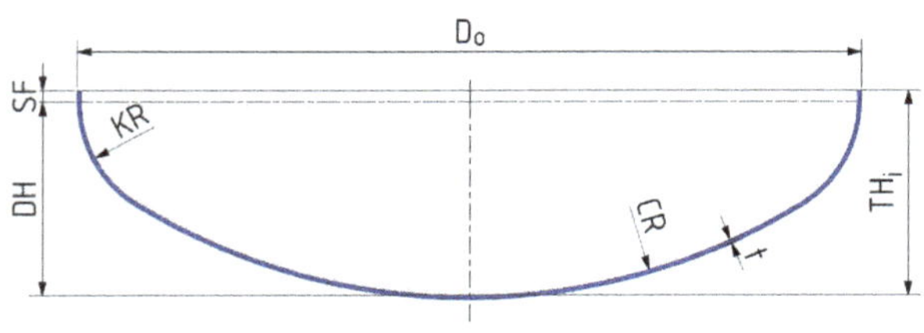

Fig. 24.3: Determining the efficiency of longitudinal and diagonal ligaments [25].

Here drums head will be designed as per ASME VIII Div. 1, which is provided below:

$$D_O = D_i + 2t$$

where D_O is external head diameter, D_i is internal head diameter and t is wall thickness

$$D_O = 48 + 2 \times 2$$

$$\boldsymbol{D_O = 45\ \text{in}}$$

↓

$$\text{CR} = 0.9D_i$$

CR is crown radius

$$\text{CR} = 0.9 \times 48$$

$$\textbf{CR = 43.2 in}$$

↓

$$\text{KR} = 0.17D_i$$

KR is knuckle radius

$$\text{KR} = 0.17 \times 48$$

$$\textbf{KR = 8.16 in}$$

↓

$$\text{DH} = 0.25D_i$$

DH is depth of dishing

$$\text{DH} = 0.25 \times 48$$

$$\textbf{DH = 12 in}$$

↓

$$\text{TH}_i = \text{SF} + \text{DH}$$

SF is straight flange height and TH_i is total internal head height

$$\text{TH}_i = 1.9865 + 12$$

$$\boldsymbol{\text{TH}_i = 13.9865\ \text{in}}$$

24.14 Mud drum selected thickness

As previously mentioned in steam drum thickness calculation, mud drum thickness calculation is given by
- Mud drum diameter = 30 in
- Mud drum thickness = 1.25 in

24.15 Mud drum ellipsoidal head

According to the mentioned formula for ellipsoidal head at steam drum head section, below data will be obtained:

$$D_O = 32.5 \text{ in} \quad CR = 27 \text{ in}$$

$$KR = 5.1 \text{ in} \quad DH = 7.5 \text{ in}$$

$$TH_i = 9.4865 \text{ in}$$

25 Bank tube average length

Bank tube average length will be calculated by the following sequence:
- Height to drilling tube inside steam drum
- Height to drilling tube inside mud drum
- Height between drum's center
- Bank tube average length

25.1 Height to drilling tube inside steam drum

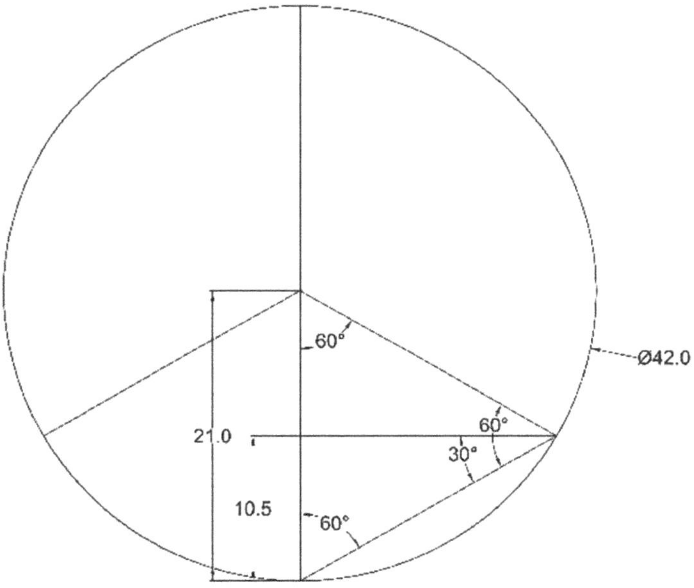

At this step, we assumed bank tube diameter and rows deep number as 2 and 14 in, respectively; then, after bank tube primary calculation then by finalizing the height between drums, tube rows deep number can be concluded:

$$\theta_{\text{total drilling}} = \theta_{\text{circumferential}} \times N_{\text{rows deep}_{\text{bank tube}}}$$

$$\theta_{\text{total drilling}} = 8^{\circ} \times 15 = 112^{\circ}$$

https://doi.org/10.1515/9783110757088-025

$$H_{\text{up to drilling steam drum}} = \frac{D_{\text{steam drum}}}{2} \times \sin\left(\frac{\text{radian}(180^\circ - \theta_{\text{total drilling}})}{2}\right)$$

$$H_{\text{up to drilling steam drum}} = \frac{48}{2} \times \sin\left(\frac{\text{radian}(180^\circ - 112^\circ)}{2}\right)$$

$$\boldsymbol{H_{\text{up to drilling steam drum}} = 13.42\ \text{in} = 340.88\ \text{mm}}$$

25.2 Height to drilling tube inside mud drum

$$H_{\text{up to drilling mud drum}} = \frac{D_{\text{mud drum}}}{2} \times \sin\left(\frac{\text{radian}(180^\circ - \theta_{\text{total drilling}})}{2}\right)$$

$$H_{\text{up to drilling mud drum}} = \frac{30}{2} \times \sin\left(\frac{\text{radian}(180^\circ - 112^\circ)}{2}\right)$$

$$\boldsymbol{H_{\text{up to drilling steam drum}} = 8.37\ \text{in} = 213.05\ \text{mm}}$$

25.3 Height between drum's center

$$H_{\text{drums CC}} = H_{\text{furnace}} + \frac{D_{\text{steam drum}}}{2} - H_{\text{up to drilling steam drum}} + \frac{D_{\text{mud drum}}}{2} - H_{\text{up to drilling mud drum}}$$

$$H_{\text{drums CC}} = 144 + \frac{48}{2} - 13.42 + \frac{30}{2} - 8.38$$

$$\boldsymbol{H_{\text{drums CC}} = 164\ \text{in}}$$

Fig. 25.1: 35 T/h, 42 barg, 420 °C water tube boiler, Morvarid Petrochemical, Assalouyeh, Iran.

25.4 Bank tube average length

$$L_{\text{aver.}_{\text{tubes}}} = H_{\text{drum CC}} - \frac{D_{\text{steam drum}} + D_{\text{mud drum}}}{2} + \frac{H_{\text{up to drilling}_{\text{steam drum}}} + H_{\text{up to drilling}_{\text{mud drum}}}}{2}$$

$$L_{\text{aver.}_{\text{tubes}}} = 163.17 - \frac{48 + 30}{2} + \frac{13.42 + 8.38}{2}$$

$$\boldsymbol{L_{\text{average}_{\text{banktubes}}} = 135.07\ \text{in}}$$

26 Bank tube heat duty prediction

The heat required to transfer from a hot side to the cold side over a unit of time is named as heat duty. First, we should calculate bank tube heat duty as per the given formulae:

Fig. 26.1: Water tube boiler (132 T/h, 42 barg, 390 °C), Basra Petrochemical, Basra, Iraq.

Bank tube water inlet temperature	345.2	F
Bank tube inlet enthalpy	317.73	Btu/lb
Bank tube steam outlet temperature	493.8	F
Bank tube outlet enthalpy	499,93	Btu/lb
Boiler output flow rate	132,277.2	lb/h
Blowdown percentage	3	%
Feed water to drum mass flow	136,245.52	lb/h

https://doi.org/10.1515/9783110757088-026

$$Q_{\text{duty}} = (h_{\text{saturated steam}} - h_{\text{inlet water}}) \times m'_{\text{Drum water @ 100\% MCR}}$$

$$Q_{\text{duty}} = (499.93 - 317.73)\,\text{Btu/lb} \times 136,245.52\,\text{lb/hr}$$

$$\mathbf{Q_{duty} = 24.82\,MMBtu/hr}$$

27 Bank tube heat surface prediction

Bank tube heat surface prediction will be calculated by the following sequence:

- Bank tube area width
- Bank tube transverse pitch
- Bank tube thickness
- Bank tube area
- Bank tube row number
- Bank tube heat surface prediction

27.1 Bank tube area width

The bank tube area can by assuming flue gas velocity and controlling gas velocity be determined.

Fig. 27.1: Water tube boiler (35 T/h, 42 barg, 420 °C), Morvarid Petrochemical, Assalouyeh, Iran.

Flue gas velocity range, 3,000–6,000 fpm [41]
Selected flue gas velocity = 5,100 fpm
Flue gas flow = 170,818.92 lb/h
Flue gas density = 0.019 lb/ft^3

https://doi.org/10.1515/9783110757088-027

$$V'_{\text{flue gas}} = \frac{170{,}818.92}{0.019} = 8{,}971{,}221\ \text{ft}^3/\text{h}$$

$$A_{\text{required}} = 8{,}971{,}221/5{,}100 = 29.32\,\text{ft}^2$$

$$W_{\text{flue gas side}} = \frac{A_{\text{required}}}{L_{\text{average}_{\text{bank tubes}}}} = \frac{29.32 \times 12}{163.17} = 2.6\,\text{ft}$$

$$W_{\text{bank tube}} = W_{\text{flue gas side}} + (N_{\text{rows deep}_{\text{bank tubes}}} - 1) \times d_{\text{tube}} = 2.6 \times 12 + (14 - 1) \times 2$$

$$\mathbf{W_{\text{bank tube}} = 57.26\ \text{in}}$$

27.2 Bank tube transverse pitch

Transverse pitch is the distance between the rows of tube, and can be obtained by the following formulae:

$$S_{\text{T}} = \frac{W_{\text{bank tube}}}{N_{\text{rows deep}_{\text{bank tubes}}}}$$

$$S_{\text{T}} = \frac{57.26}{14}$$

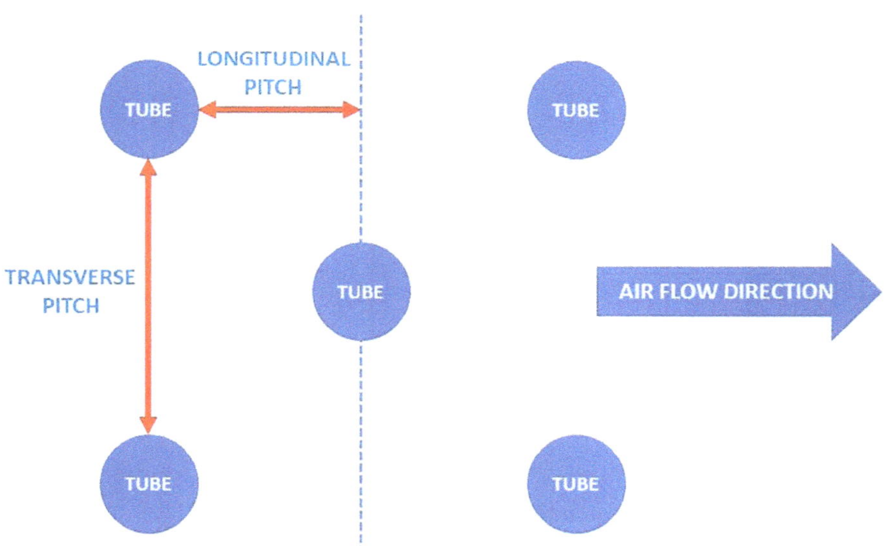

Fig. 27.2: Transverse and longitudinal pattern.

$$\mathbf{S_T = 4.09\ in}$$

As you can see, the assumption for transvers pitch is corrected, but needs to be checked by bank tube performance and draft loss later.

27.3 Bank tube thickness

$$t_{\min} = \frac{PD}{2S + P} + 0.005D + e$$

where t is minimum required tube thickness, inch; P is boiler design pressure, psig; D is tube diameter, inch; S is the maximum allowable stress for A213-T11, psi; and e is the thickness factor for expanded tube ends as per ASME 1, PG-27.4

$$t_{\min} = \frac{48.16 \times 14.2233 \times 2}{2 \times 12{,}400 + 48.16 \times 14.2233} + 0.005 \times 2 + 0.04$$

$$t_{\min} = 0.103\ \text{in}$$

As per tube manufacture catalog for 2-in diameter, they produce tube with a thickness of 0.105 in. So,

$$\mathbf{t_{selected} = 0.105\ in}$$

27.4 Bank tube area

$$A_{\text{tube}} = \frac{\pi}{4} \times \left[\frac{(d_{\text{tube}} - 2 \times t)}{12}\right]^2$$

$$A_{\text{tube}} = \frac{\pi}{4} \times \left[\frac{(2 - 2 \times 0.105)}{12}\right]^2$$

$$\mathbf{A_{tube} = 0.0174\ ft^2}$$

27.5 Bank tube row number

Circulation ratio comes from portion of the passes flows in the system and the amount of generated steam from steam drum, which typically for D-type water tube boiler is 40.

Here, by calculation at following steps, we can find circulation ratio as 40 which, at this step, we should find simulation of bank tube arrangement and after that we can calculate the circulation ratio and then come back to recalculate bank tube arrangement.

Fig. 27.3: Water tube boiler (35 T/h, 42 barg, 420 °C), Morvarid Petrochemical, Assalouyeh, Iran.

Water velocity ranges inside boiler tubes, 70–700 fpm [41]
Selected water velocity = 200 fpm
Feed water to drum flow rate = 136,245.52 lb/h
Saturated water density = 23.98 lb/ft^3
Bank tube diameter = 2 in
Bank tube thickness = 0.105 in

$$V'_{\text{boiler circulation water}} = \frac{m'_{\text{fw to drum}} \times \text{CR}}{\rho_{\text{saturated water}}}$$

$$V'_{\text{boiler circulation water}} = \frac{136{,}245.52 \times 40}{26.63}$$

$$\mathbf{V'_{\text{boiler circulation water}} = 204{,}613.12\ ft^3/h}$$

↓

$$N_{\text{total bank tubes}} = \text{Round up}\left(\frac{V'_{\text{boiler circulation water}}}{V_{\text{selected water}} \times A_{\text{tube}} \times 60}\right)$$

$$N_{\text{total}_{\text{bank tubes}}} = \text{Round up}\left(\frac{204{,}613.12}{200 \times 0.0174 \times 60}\right)$$

$$\boldsymbol{N_{\text{total}_{\text{bank tubes}}} = 976 \text{ No.}}$$

$$N_{\text{rows}_{\text{bank tubes}}} = \text{Round up}\left(\frac{N_{\text{total}_{\text{bank tubes}}}}{N_{\text{rows deep}_{\text{bank tubes}}}}\right)$$

$$N_{\text{rows}_{\text{bank tubes}}} = \text{Round up}\left(\frac{976}{14}\right)$$

$$\boldsymbol{N_{\text{rows}_{\text{bank tubes}}} = 70 \text{ number}}$$

27.6 Bank tube heat surface prediction

$$A_{\text{bank tube}} = \pi \times \frac{D_{\text{tube}}}{12} \times \text{tube row no.} \times (\text{tube row deep no.} - 1) \times L_{\text{tube}}$$

$$A_{\text{bank tube}} = \pi \times \frac{2}{12} \times 70 \times (14 - 1) \times \frac{135.07}{12}$$

$$\boldsymbol{A_{\text{bank tube}} = 5{,}360.46 \text{ft}^2 = 498 \text{ m}^2}$$

28 Steam drum outlet steam temperature

Steam drum outlet steam temperature will be calculated by the following sequence:
- Bank tube inlet flue gas properties
- Bank tube flue gas gas mass velocity
- Bank tube flue gas heat transfer coefficient
- Bank tube package performance prediction
- Bank tube bundle flue gas outlet temperature
- Bank tube bundle flue gas average temperature
- Steam drum primary outlet steam temperature
- Drum each tube circulation water
- Inside gas film resistance
- Metal resistance
- Outside gas film resistance
- Heat flux
- Temperature drop across the gas film
- Temperature drop across the tube metal
- Temperature drop across the steam film
- Steam drum outlet steam temperature

28.1 Bank tube inlet flue gas properties

Flue gas properties at inlet temperature can be found from Tab. 28.1:

$$T_{\text{gas}_{\text{outlet}}} = 1{,}750.96\ ^\circ\text{F}$$

$$\mu = 0.10965\ \text{lb/ft h}$$

$$C_p = 0.32491\ \text{Btu/lb F}$$

$$K = 0.04648\ \text{Btu/ft h F}$$

Tab. 28.1: Flue gas product properties.

Temperature (°F)	C_P	μ	K
2,000	0,3326	0,1174	0,0511
1,900	0,329,525	0,1143	0,04925
1,800	0,32,645	0,1112	0,0474
1,700	0,323,375	0,1081	0,04555
1,600	0,3203	0,105	0,0437

https://doi.org/10.1515/9783110757088-028

28.2 Bank tube flue gas gas mass velocity

Gas mass velocity is mass flow rate across a unit area vertical to the direction of the velocity vector and can be found as follows [5]:

$$G = 12\frac{W'}{N_{\text{row deep}} \times L_{\text{tube}} \times (S_T - d)}$$

$$G = 12 \times \frac{170{,}818.92}{14 \times \frac{135.07}{12} \times (4.09 - 2)}$$

$$\mathbf{G = 6{,}224.79\,lb/ft^2\,h}$$

28.3 Bank tube flue gas heat transfer coefficient

$$h_C = 0.9 \times \frac{G^{0.6}}{D_{\text{tube}}^{\;0.4}} \times \frac{K^{0.67} \times C_P^{\;0.33}}{\mu^{0.27}}$$

$$h_C = 0.9 \times \frac{6,224.79^{0.6}}{2^{0.4}} \times \frac{0.04648^{0.67} \times 0.32491^{0.33}}{0.10965^{0.27}}$$

$$\mathbf{h_C = 20.67\,Btu/ft^2\,h\,F}$$

28.4 Bank tube package performance prediction

Duty and exit temperatures can be predicted by number of transfer units (NTU) method. Fundamentally, the duty Q is given by [5]

$$Q = \varepsilon C_{\min}\left(T_{\text{gas}_{\text{ inlet}}} - T_{\text{steam}_{\text{ inlet}}}\right)$$

where ε depends on the type of flow, which can be counter flow or parallel flow or crossflow. In bank tube, usually when both streams are unmixed, a counter flow arrangement for flue gas and cross flow arrangement for saturated water is adopted. ε for these are given by

$$\varepsilon_{\text{flue gas}} = \frac{1 - \exp[-\text{NTU} \times (1 - C)]}{1 - C \times \exp[-\text{NTU} \times (1 - C)]}$$

$$\varepsilon_{\text{sat water}} = 1 - \exp\left\{C \times \text{NTU}^{0.22} \times \left[\exp\left(-C \times \text{NTU}^{0.78}\right) - 1\right]\right\}$$

where

$$\text{NTU} = \frac{\text{UA}}{C_{\min}}$$

$$C = \frac{C_{\min}}{C_{\max}}$$

$$C_{\min} = \left(W'C_p\right)_{\min}$$

$$C_{\max} = \left(W'C_p\right)_{\max}$$

At our design case,

Saturated vaporspecific heat @ constant volume = 0.604 Btu/lb F
Inlet flue gas heat capacity @ constant pressure = 0.3249 Btu/lb F

$$W'_{\text{feed water to drum}} = 136{,}245.52\,\text{lb/h}$$

$$W'_{\text{flue gas}} = 170{,}818.92\,\text{lb/h}$$

↓

$$C_{\max} = 0.604 \times 136{,}245.52 = 82{,}315.20$$

$$C_{\min} = 0.3249 \times 170{,}818.92 \times \left(1 - \frac{2}{100}\right) = 54{,}391.18$$

$$C = \frac{54{,}391.18}{82{,}315.20} = 0.66$$

Since gas temperature is low, nonluminous heat transfer is low and can be neglected.

↓

$$\text{NTU} = \frac{20.67 \times 5{,}360.46}{54{,}391.18} = 2.037$$

↓

$$\varepsilon = \frac{1 - \exp[-2.037 \times (1 - 0.66)]}{1 - 0.66 \times \exp[-2.037 \times (1 - 0.66)]} = 0.745$$

↓

$$Q = 0.745 \times 54{,}391.18 \times (1{,}750.96 - 345.2)$$

$$\mathbf{Q = 57.036\ MMBtu/h}$$

28.5 Bank tube bundle flue gas outlet temperature

$$T_{\text{gas outlet}} = T_{\text{gas inlet}} - \frac{Q}{C_p \times W'_{\text{gas}} \times (1 - \text{heat loss\%})}$$

$$T_{\text{gas outlet}} = 1{,}750.96 - \frac{57.036 \times 10^6}{0.3249 \times 170{,}818.92 \times \left(1 - \left(\frac{2}{100}\right)\right)}$$

$$\mathbf{T_{\text{gas outlet}} = 662.63\ ^\circ F}$$

Note: Drum steam outlet depends on heat surface and technical velocity on bank tube so tube quantity needs to be checked by drum outlet temperature

Note: Tube row deep need to be checked by permissible velocity on furnace (bank tube draft loss calculation)

28.6 Bank tube bundle flue gas average temperature

$$T_{\text{average}} = \frac{662.63 + 1{,}750.96}{2}$$

$$\mathbf{T_{\text{average}} = 1{,}206.79\ ^\circ F}$$

28.7 Steam drum primary outlet steam temperature

$$T_{\text{primary steam outlet}} = T_{\text{steam inlet}} + \frac{Q}{C_p \times W'_{\text{steam}}}$$

$$T_{\text{primary steam outlet}} = 345.2 + \frac{57.036 \times 10^6}{0.604 \times 136{,}245.52}$$

$$\mathbf{T_{\text{primary steam outlet}} = 1{,}038.1\ ^\circ F}$$

28.8 Drum each tube circulation water

$$m'_{\text{each tube circulation water}} = \frac{m'_{\text{drum circulation water}}}{N_{\text{bank tubes}}}$$

$$m'_{\text{each tube circulation water}} = \frac{5{,}499{,}820.64}{976}$$

$$\mathbf{m'_{\text{each tube circulation water}} = 5{,}583.83\ lb/h}$$

28.9 Inside gas film resistance

$$R_{\text{Inside gas film}} = \frac{1}{h_{\text{tube inside coeff.}}}$$

$$h_{\text{tube inside coeff.}} = \frac{2.44 \times m'_{\text{each tube circulation water}}{}^{0.8} \times C}{(d_{\text{tube}} - t_{\text{tube}})^{1.8}}$$

$$P_{\text{sat.}} = 643.69 \text{ psia}$$

$$T_{\text{sat.}} = 493.8 \ ^{\circ}\text{F}$$

$$C = 0.36$$

Tab. 28.2: Factor *C* for steam [5].

Temperature (°F)	Drum saturation pressure (psia)				
	100	**200**	**500**	**1,000**	**2,000**
400	0.2716	0.3059	–	–	–
500	0.2737	0.2909	0.3595	–	–
600	0.2813	0.2896	0.3228	0.413	–
700	0.2917	0.2965	0.3161	0.3586	0.5206
800	0.3050	0.3090	0.3206	0.3453	0.4214
900	0.3161	0.3197	0.3277	0.3477	0.3946
1,000	0.3276	0.3302	0.3392	0.3531	0.386

$$h_{\text{tube inside coeff.}} = \frac{2.44 \times 5,583.83^{0.8} \times 0.36}{(2 - 0.105)^{1.8}}$$

$$\boldsymbol{h_{\text{tube inside coefficient}} = 276.40\ \text{Btu/ft}^2\ \text{h F}}$$

↓

$$R_{\text{Inside gas film}} = \frac{1}{h_{\text{tube inside coeff.}}}$$

$$R_{\text{Inside gas film}} = \frac{1}{276.40}$$

$$\boldsymbol{R_{\text{Inside gas film}} = 0.0036\ \ \text{ft}^2\ \text{h F/Btu}}$$

28.10 Metal resistance

$$R_{\text{m}} = \frac{d_{\text{tube}}}{24K_{\text{m}}} \times \ln\frac{d_{\text{tube}}}{(d_{\text{tube}} - t_{\text{tube}})}$$

↓

$$\boldsymbol{T_{\text{average flue gas}} = 1,206.79\ ^{\circ}\text{F}}$$

↓

Thermal conductivity of metals

$$K_{\text{m}} = 27\ \ \text{Btu/ft h F}$$

$$R_{\text{m}} = \frac{2}{24 \times 27} \times \ln\frac{2}{(2 - 0.105)}$$

$$\boldsymbol{R_{\text{m}} = 0.00016\ \text{ft}^2\ \text{h F/Btu}}$$

28.11 Outside gas film resistance

$$R_{\text{o}} = \frac{1}{h_{\text{o}}}$$

$$R_{\text{o}} = \frac{1}{20.67}$$

$$\boldsymbol{R_{\text{o}} = 0.0483\ \ \text{ft}^2\ \text{h F/Btu}}$$

28.12 Heat flux

$$Q_{\text{flux}} = \frac{\left(T_{\text{average flue gas}} - T_{\text{primary steam outlet.}}\right)}{\left(R_{\text{Inside gas film}} + R_{\text{m}} + R_{\text{o}}\right)}$$

$$Q_{\text{flux}} = \frac{(1{,}206.79 - 1{,}038.1)}{(0.0036 + 0.00016 + 0.0495)}$$

$$\boldsymbol{Q_{\text{flux}} = 3{,}234.55\ \text{Btu/ft}^2\ \text{h F}}$$

28.13 Temperature drop across the gas film

$$\Delta T_{\text{across the gas film}} = Q_{\text{flux}} \times R_{\text{Inside gas film}}$$

$$\Delta T_{\text{across the gas film}} = 3{,}234.55 \times 0.0036$$

$$\boldsymbol{\Delta T_{\text{across the gas film}} = 156.45\ ^{\circ}\text{F}}$$

28.14 Temperature drop across the tube metal

$$\Delta T_{\text{across the tube metal}} = Q_{\text{flux}} \times R_{\text{m}}$$

$$\Delta T_{\text{across the tube metal}} = 3{,}234.55 \times 0.00016$$

$$\boldsymbol{\Delta T_{\text{across the tube metal}} = 0.53\ ^{\circ}\text{F}}$$

28.15 Temperature drop across the steam film

$$\Delta T_{\text{across the steam film}} = Q_{\text{flux}} \times R_{\text{o}}$$

$$\Delta T_{\text{across the steam film}} = 3{,}234.55 \times 0.0495$$

$$\boldsymbol{\Delta T_{\text{across the steam film}} = 11.7\ ^{\circ}\text{F}}$$

28.16 Steam drum outlet steam temperature

$$T_{\text{outlet steam}} = T_{\text{outlet flue gas}} - \Delta T_{\text{across the gas film}} - \Delta T_{\text{across the tube metal}} - \Delta T_{\text{across the steam film}}$$

$$T_{\text{outlet steam}} = 662.63 - 156.45 - 0.53 - 11.7$$

$$\boldsymbol{T_{\text{outlet steam}} = 493.94\ ^{\circ}\text{F}}$$

29 Bank tube bundle flue gas draft pressure drop

Bank tube bundle flue gas draft pressure drop will be calculated by the below sequence:

- Bank tube bundle flue gas density
- Bank tube bundle flue gas average temperature viscosity
- Bank tube bundle flue gas Reynolds number
- Bank tube bundle flue gas friction factor
- Bank tube bundle flue gas draft pressure drop

29.1 Bank tube bundle flue gas density

$$\rho_{\text{flue gas}} = 492 \times 29.24 \times \frac{15.53 - 1.88 \times 0.0361273}{359 \times (460 + 1{,}206.8 \times 14.7)}$$

$$\rho_{\textbf{flue gas}} = \mathbf{0.0253\ lb/cu.ft}$$

29.2 Bank tube bundle flue gas average temperature viscosity

Flue gas properties at average temperature can be found from the below equation:

$$\mu = 0.0926\ \text{lb/ft h}$$

29.3 Bank tube bundle flue gas Reynolds number

$$\text{Re} = \frac{Gd_{\text{Tube}}}{12\mu}$$

$$\text{Re} = \frac{6{,}224.79 \times 2}{12 \times 0.0926}$$

$$\textbf{Re} = \mathbf{11{,}206.75}$$

https://doi.org/10.1515/9783110757088-029

29.4 Bank tube bundle flue gas friction factor

The friction factor of turbulent flow is given by [5]

$$\text{fr} = \text{Re}^{-0.15} \times \left(0.044 + \frac{0.08 \times \frac{S_L}{d_{\text{tube}}}}{\left(\frac{S_T}{d_{\text{tube}}} - 1\right)^{0.43 + 1.13 \times \frac{d_{\text{tube}}}{S_L}}} \right)$$

$$\text{fr} = 11,206.75^{-0.15} \times \left(0.044 + \frac{0.08 \times \frac{4.03125}{2}}{\left(\frac{4.09}{2} - 1\right)^{0.43 + 1.13 \times \frac{2}{4.03125}}} \right)$$

$$\mathbf{fr = 0.049}$$

29.5 Bank tube bundle flue gas draft pressure drop

Flue gas draft pressure drop in bank tube bundle can be obtained by the following formulae [5]:

$$\Delta P_{\text{bank tube bundle}} = 9.3 \times 10^{-10} \times G^2 \times \text{fr} \times \frac{N_{\text{tube row deep}}}{\rho}$$

$$\Delta P_{\text{bank tube bundle}} = 9.3 \times 10^{-10} \times 6,224.79^2 \times 0.049 \times \frac{14}{0.0253}$$

$$\Delta P_{\text{bank tube bundle}} = 0.98 \text{ in WG} = 24.83 \text{ mm WG}$$

30 Bank tube duct flue gas draft pressure drop

Bank tube duct flue gas draft pressure drop will be calculated by the following sequence:

- Bank tube flue gas volumetric flow rate
- Bank tube flue gas velocity
- Bank tube bundle exit flue gas density
- Bank tube bundle exit flue gas volumetric flow rate
- Bank tube bundle exit width
- Head loss due to change in bank tube duct contraction
- Head loss due to bank tube duct sudden contraction
- Bank tube duct flue gas draft pressure drop

30.1 Bank tube flue gas volumetric flow rate

$$W'_{\text{bank tube flue gas}} = \frac{170,818.92}{0.0253}$$

$$\boldsymbol{W'_{\text{bank tube flue gas}} = 6,757,653 \ \text{cu.ft/h}}$$

30.2 Bank tube flue gas velocity

$$V_{\text{bank tube flue gas}} = \frac{G}{60\rho}$$

$$V_{\text{bank tube flue gas}} = \frac{6,224.79}{60 \times 0.0253}$$

$$\boldsymbol{V_{\text{bank tube flue gas}} = 4,104.24 \ \text{fpm}}$$

30.3 Bank tube bundle exit flue gas density

$$\rho_{\text{flue gas}} = 492 \times 29.24 \times \frac{15.46 - 0.98 \times 0.0361273}{359 \times (460 + 662.63 \times 14.7)}$$

$$\boldsymbol{\rho_{\text{flue gas}} = 0.0374 \ \text{lb/cu.ft}}$$

https://doi.org/10.1515/9783110757088-030

30.4 Bank tube bundle exit flue gas volumetric flow rate

$$W'_{\text{exit flue gas}} = \frac{170,818.92}{0.0374}$$

$$W'_{\text{exit flue gas}} = 4,561,885.51 \text{ cu.ft/h}$$

30.5 Bank tube bundle exit width

Flue gas velocity inside boiler passes should be according to velocity range table. It should be in the range of 3,000 to 6,000 fpm and selected velocity will be 3,971.7 fpm.

$$W = \frac{V'_{\text{exit flue gas}}}{v'_{\text{exit flue gas}} \times L_{\text{average bank tubes}}}$$

$$W = \frac{4,561,885.51}{\frac{135.07}{12} \times 4,104.25 \times 60}$$

$$W = 1.65 \text{ ft}$$

30.6 Head loss due to change in bank tube duct contraction

Head loss can be obtaining by using the below two factors to find contraction correction factor [16].

$$\frac{H}{W1} = \frac{135.07}{57.26} = 2.36$$

$$\frac{W0}{W1} = \frac{1.65 \times 12}{57.26} = 0.34$$

SR3-1 Elbow, 90 Degree, Variable Inlet/Outlet Areas, Supply Air Systems

	C_o Values						
	W_o/W_1						
H/W_1	0.6	0.8	1.0	1.2	1.4	1.6	2.0
0.25	0.63	0.92	1.24	1.64	2.14	2.71	4.24
1.00	0.61	0.87	1.15	1.47	1.86	2.30	3.36
4.00	0.53	0.70	0.90	1.17	1.49	1.84	2.64
100.	0.54	0.67	0.79	0.99	1.23	1.54	2.20

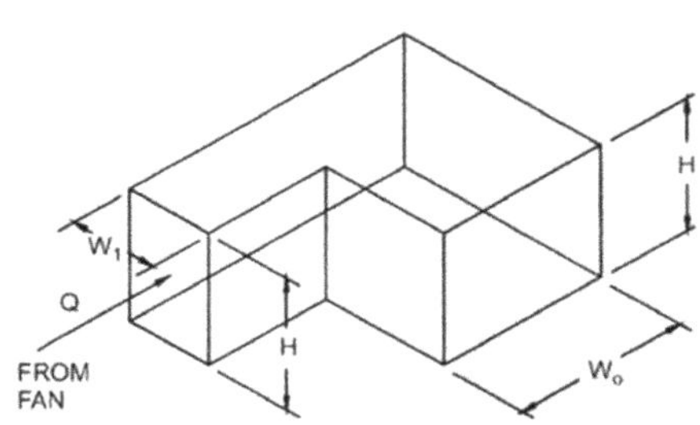

Fig. 30.1: 132 T/h, 42 barg, 390 °C water tube boiler, Basra Petrochemical, Basra, Iraq.

Fig. 30.2: Duct correction factor [16].

$$f_{\text{contraction}} = 0.57$$

$$h_{\text{loss contraction}} = \frac{1}{2} \times f_{\text{contraction}} \times \rho_{\text{flue gas}} \times \left(\frac{V'_{\text{exit flue gas}}}{60}\right)^2$$

$$h_{\text{loss contraction}} = \frac{1}{2 \times 32} \times 0.57 \times 0.0374 \times \left(\frac{4,104.25}{60}\right)^2$$

$$\mathbf{h_{\text{loss contraction}} = 1.55\ ft}$$

30.7 Head loss due to bank tube duct sudden contraction

$$f_{\text{sudden contraction}} = 0.57$$

$$h_{\text{loss sudden contraction}} = \frac{1}{2 \times 32} \times f_{\text{sudden contraction}} \times \rho_{\text{flue gas}} \times \left(\frac{V'_{\text{exit flue gas}}}{60}\right)^2$$

$$h_{\text{loss sudden contraction}} = \frac{1}{2 \times 32} \times 0.57 \times 0.0374 \times \left(\frac{4,104.25}{60}\right)^2$$

$$\mathbf{h_{loss\ sudden\ contraction} = 1.55\ ft}$$

30.8 Bank tube duct flue gas draft pressure drop

$$\Delta P_{\text{bank tube duct}} = h_{\text{loss contraction}} + h_{\text{loss sudden contraction}}$$

$$\Delta P_{\text{bank tube duct}} = \frac{(1.55 + 1.55) \times 12}{62.4}$$

$$\mathbf{\Delta P_{bank\ tube\ duct} = 0.596\ in\ WG = 15.15\ mm\ WG}$$

31 Bank tube total flue gas draft pressure drop

$$\Delta P_{\text{bank tube}} = \Delta P_{\text{bank tube duct}} + \Delta P_{\text{bank tube bundle}}$$

$$\Delta P_{\text{bank tube}} = 0.596 + 0.98$$

$$\boldsymbol{\Delta P_{\text{bank tube}} = 1.57 \text{ in WG} = 39.98 \text{ mm WG}}$$

https://doi.org/10.1515/9783110757088-031

32 Bank tube area length

$$L_{\text{bank tube area}} = S_L \times (N_{\text{tube row}} + 2)$$

$$L_{\text{bank tube area}} = 4.03125 \times (70 + 2)$$

$$\mathbf{L_{\text{bank tube area}} = 290.3\ in}$$

Fig. 32.1: Water tube boiler (132 T/h, 42 barg, 390 °C), Basra Petrochemical, Basra, Iraq.

https://doi.org/10.1515/9783110757088-032

33 Furnace area length

$$L_{\text{furnace}} = L_{\text{bank tube area}} + L_{\text{superheater}} + L_{\text{turning lane}}$$

$$\downarrow$$

$$L_{\text{superheater}} = N_{\text{superheater tube row}} \times S_{\text{L}}$$

$$L_{\text{superheater}} = 19 \times 4.03125$$

$$\boldsymbol{L_{\text{superheater}} = 73\ \text{in}}$$

$$\downarrow$$

$$L_{\text{furnace}} = 290.3 + 73 + 40$$

$$\boldsymbol{L_{\text{furnace}} = 402.8\ \text{in}}$$

https://doi.org/10.1515/9783110757088-033

34 Boiler exit duct flue gas draft pressure drop

Boiler exit duct flue gas draft pressure drop will be calculated by the following sequence:

- Boiler exit duct dimension
- Boiler exit duct flue gas density
- Boiler exit duct equivalent diameter
- Boiler exit duct equivalent length
- Boiler exit duct flue gas Reynolds number
- Boiler exit duct flue gas friction factor
- Boiler exit duct flue gas draft pressure drop

34.1 Boiler exit duct dimension

Height and width of duct in this design will be same as boiler exit dimension and will be 135.07 in and 1.65 ft, respectively.

34.2 Boiler exit duct flue gas density

$$\rho_{\text{flue gas}} = 492 \times 29.24 \times \frac{15.46 - 1.57 \times 0.0361273}{359 \times (460 + 662.63 \times 14.7)}$$

$$\boldsymbol{\rho_{\text{flue gas}} = 0.0374\ \text{lb/cu.ft}}$$

34.3 Boiler exit duct equivalent diameter

$$D_{\text{eq.}} = 1.3 \times \frac{(W \times H)^{0.625}}{(W + H)^{0.25}}$$

$$D_{\text{eq.}} = 1.3 \times \frac{(1.65 \times 12 \times 135.07)^{0.625}}{(1.65 \times 12 + 135.07)^{0.25}}$$

$$\boldsymbol{D_{\text{eq.}} = 51.03\ \text{in}}$$

https://doi.org/10.1515/9783110757088-034

34.4 Boiler exit duct equivalent length

$$L_{eq.} = L_{duct} + L_{elbow\ equal}$$

$$L_{eq.} = 10 + 25$$

$$\boldsymbol{L_{eq.} = 35\,\mathrm{ft}}$$

34.5 Boiler exit duct flue gas Reynolds number

$$\mathrm{Re} = 15.2 \times \frac{W'}{\mu \times D_{eq.}}$$

$$\mathrm{Re} = 15.2 \times \frac{170,818.92}{0.0706 \times 51.03}$$

$$\mathbf{Re = 720,458.29}$$

34.6 Boiler exit duct flue gas friction factor

$$\mathrm{fr} = \frac{0.316}{\mathrm{Re}^{0.25}}$$

$$\mathrm{fr} = \frac{0.316}{720,458.29^{0.25}}$$

$$\mathbf{fr = 0.011}$$

34.7 Boiler exit duct flue gas draft pressure drop

$$\Delta P_{air} = 9.3 \times 10^{-5} \times \mathrm{f} \times W'^{2} \times \frac{L_e}{\rho \times d_i^5}$$

$$\Delta P_{air} = 9.3 \times 10^{-5} \times 0.011 \times 170,818.92^2 \times \frac{35}{0.0374 \times 51.03^5}$$

$$\boldsymbol{\Delta P_{air} = 0.0796\ \ \mathrm{in\,WG} = 2.02\,\mathrm{mm\,WG}}$$

35 Bank tube bundle water pressure drop

Bank tube bundle water pressure drop will be calculated by the following sequence:

- Bank tube bundle water flow
- Bank tube bundle water flow Reynolds number
- Bank tube bundle water flow friction factor
- Bank tube bundle water pressure drop

35.1 Bank tube bundle water flow

Water circulating inside bank tubes is equal to inlet water multiplied with circulation ratio and blowdown rate which here we consider as 40% and 3%, and at next steps, we will calculate the circulation ratio.

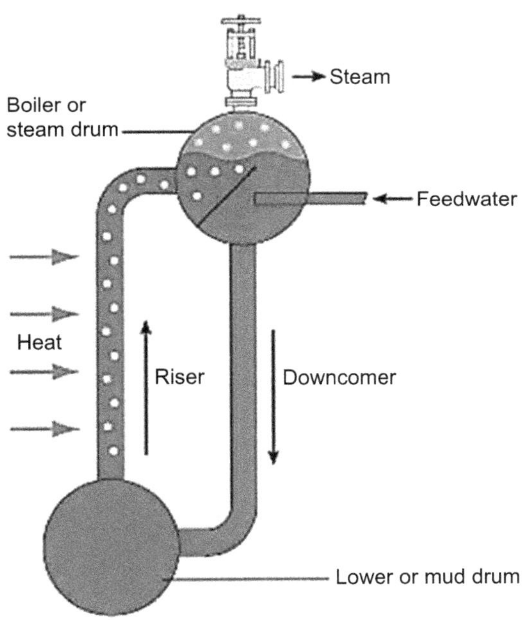

Fig. 35.1: Steam and water stream at boiler.

https://doi.org/10.1515/9783110757088-035

35.2 Bank tube bundle water flow Reynolds number

$$\mathrm{Re} = 15.2 \times \frac{\dfrac{W'}{N_{\text{rows deep}}}}{\mu \times d_{\text{i}}}$$

$$\mathrm{Re} = 15.2 \times \frac{\dfrac{136,245.516 \times 40}{70 \times 14}}{0.249552 \times (2 - 2 \times 0.105)}$$

$$\mathbf{Re = 189.228}$$

35.3 Bank tube bundle water flow friction factor

The friction factor of turbulent flow of water inside the tube is given by

$$\mathrm{fr} = \frac{0.316}{\mathrm{Re}^{0.25}}$$

$$\mathrm{fr} = \frac{0.316}{189.228^{0.25}}$$

$$\mathbf{fr = 0.015}$$

35.4 Bank tube bundle water pressure drop

$$\Delta P_{\text{g}} = 3.36 \times 10^{-6} \times \mathrm{fr} \times L_{\text{e}} \times \frac{\left(\dfrac{W'}{N_{\text{tube row}}}\right)^2}{\rho \times (D - t)^5}$$

$$\Delta P_{\text{g}} = 3.36 \times 10^{-6} \times 0.015 \times 11.26 \times \frac{\left(\dfrac{136,245.516 \times 40}{70 \times 14}\right)^2}{\dfrac{1}{0.02} \times (2 - 0.12)^5}$$

$$\mathbf{\Delta P_{g} = 0.014\ psig = 0.101\ kpag}$$

36 Front and rear wall headers sizing

Front and rear wall headers sizing will be calculated by the following sequence:
- Front and rear wall headers diameter
- Front and rear wall headers thickness

36.1 Front and rear wall headers diameter

$$V'_{\text{front andrear wall}} = \%_{\text{front and real wallflow rate}} \times V'_{\text{boiler circulation water}} \times 10\%$$

$$V'_{\text{front and rear wall}} = \frac{2 \times N_{\text{furnace width tangent tube}}}{N_{\text{furnace tube}}} \times V'_{\text{boiler circulation water}} \times 10\%$$

↓

$$N_{\text{furnace width tangent tube}} = 51$$

$$N_{\text{furnace tube}} = 413$$

↓

$$V'_{\text{front and rear wall}} = \frac{2 \times 51}{413} \times 204,613.12 \times 10\%$$

$$\boldsymbol{V'_{\text{front and rear wall}} = 5,053.40 \quad \text{ft}^3/\text{h}}$$

↓

$$A_{\text{required}} = \frac{V'_{\text{front and rear wall}}}{v'_{\text{inside header}}}$$

$$A_{\text{required}} = \frac{5,053.40}{500}$$

$$\boldsymbol{A_{\text{required}} = 0.168\ \text{ft}^2}$$

↓

https://doi.org/10.1515/9783110757088-036

$$D_{\text{required}_{\text{header}}} = \sqrt{\frac{4 \times A_{\text{required}}}{\pi}}$$

$$D_{\text{required}_{\text{header}}} = \sqrt{\frac{4 \times 0.168 \times 12^2}{\pi}}$$

$$D_{\text{required}_{\text{header}}} = 5.55\ \text{in}$$

$$\boldsymbol{D_{\text{selected}_{\text{header}}} = 6\ \text{in}}$$

$$V'_{\text{front and rear wall}} = \frac{60 \times V'_{\text{front and rear wall}}}{\pi/4 \times \left(\frac{D_{\text{required header}}}{12}\right)^2}$$

$$V'_{\text{front and rear wall}} = \frac{60 \times 5,053.40}{\pi/4 \times \left(\frac{6}{12}\right)^2}$$

$$\boldsymbol{V'_{\text{front and rearwall}} = 429.16\ \ \text{ft/m}}$$

If boiler circulation water range inside boiler is 7–700 ft/m, then the selected diameter is OK.

36.2 Front and rear wall headers thickness

$$t_{\text{m}} = \frac{PD}{2 \times (68.947 \times S \times E + P \times Y)} + A$$

where t is the minimum required header thickness, inch; P is header design pressure, barg; D is header diameter, mm; S is the maximum allowable stress for A106-Gr.B, Ksi; E is joint efficiency; Y is temperature coefficient; A is corrosion allowance, mm.

$$t_{\text{m}} = \frac{48.17 \times 6 \times 25.4}{2 \times (68.947 \times 15 \times 0.85 + 48.17 \times 0.4)} + 3$$

$$t_{\text{m}} = 7.08\text{mm} = 0.27\ \text{in}$$

Minimum required thickness for calculation will be 0.27 in, for which, from pipe manufacturer catalog, we can select 0.28 in thickness or schedule 40 for 6 in diameter header.

Fig. 36.1: Water tube boiler (132 T/h, 42 barg, 390 °C), Basra Petrochemical, Basra, Iraq.

37 Steam drum to superheater connection header sizing

Steam drum to superheater connection header sizing will be calculated by the following sequence:

- Steam drum to superheater pipe diameter
- Steam drum to superheater pipe thickness by pressure
- Steam drum to superheater pipe length
- Steam drum to superheater pipe steam flow Reynolds number
- Steam drum to superheater pipe steam flow friction factor
- Steam drum to superheater pipe steam flow pressure drop

37.1 Steam drum to superheater pipe diameter

Steam velocity range, 61–76.2 m/s [41]
Selected steam velocity = 61 m/s
Saturated Steam flow = 60/3.6 = 16.66 kg/s
Saturated steam specific volume = 0.044 m^3/kg
Superheater steam volumetric flow rate = 0.737 m^3/s
Area required = 0.737/61 = 0.012 m^2
Required header diameter = 4.88 in

Fig. 37.1: Water tube boiler (132 T/h, 42 barg, 390 °C), Basra Petrochemical, Basra, Iraq.

https://doi.org/10.1515/9783110757088-037

Selected header diameter = 6 in

$$\textit{Nominal steam velocity} = \frac{0.737}{\pi/4 \times \left(\frac{6 \times 1,000}{25.4}\right)^2} = 40.46\,\mathrm{m/s}$$

Then, diameter selection is fine.

37.2 Steam drum to superheater pipe thickness by pressure

$$t_{\min_1} = \frac{PD}{2 \times (68.947 \times S \times E + P \times Y)} + A$$

where t is the minimum required header thickness, inch; P is the header design pressure, barg; D is header diameter, mm; S is the maximum allowable stress for A106-Gr. B, Ksi; E is joint efficiency; Y is temperature coefficient; A is corrosion allowance, mm

$$t_{\min_1} = \frac{48.17 \times 4 \times 25.4}{2 \times (68.947 \times 15 \times 0.85 + 48.17 \times 0.4)} + 3$$

$$t_{\min_1} = 7.08\mathrm{mm} = 0.278\mathrm{in}$$

Minimum required thickness for calculation will be 0.278 in which we can obtain from the pipe manufacturer catalog. We can select 0.28 in thickness or schedule 40 for 6 in diameter header.

37.3 Steam drum to superheater pipe length

$$L_{\text{connecting drum to superheater}} = L_{\text{straight pipe}} + N_{\text{elbow}} \times L_{\text{elbow equivalent}}$$

Tab. 37.1: Equivalent length for valves and fitting [5].

Pipe inner diameter (in)	Gate valve, fully open	Swing check valve, fully open	Globe valve, fully open	90° elbow
1	0.7	8.7	30	2.6
2	1.4	17.2	60	5.2
3	2	25.5	87	7.7
4	2.7	33.5	114	10
6	4	50.5	172	15.2
8	5.3	33	225	20

$$L_{\text{connecting drum to superheater}} = 32 + 3 \times 15.2$$

$$\boldsymbol{L_{\text{connecting drum to superheater}} = 77.6\text{ft}}$$

37.4 Steam drum to superheater pipe steam flow Reynolds number

$$\text{Re} = 15.2 \times \frac{W'}{\mu \times d_{\text{i}}}$$

$$\text{Re} = 15.2 \times \frac{132,277.2}{0.249552 \times (6 - 2 \times 0.28)}$$

$$\mathbf{Re = 1,481,046.3}$$

37.5 Steam drum to superheater pipe steam flow friction factor

The friction factor of turbulent flow of steam inside the tube is given by:

$$\text{fr} = \frac{0.316}{\text{Re}^{0.25}}$$

$$\text{fr} = \frac{0.316}{1,481,046.3^{0.25}}$$

$$\text{fr} = 0.009$$

37.6 Steam drum to superheater pipe steam flow pressure drop

$$\Delta P_{\text{g}} = 3.36 \times 10^{-6} \times \text{fr} \times L_{\text{connecting drum to superheater}} \times \frac{W'^2}{\rho \times (D - t)^5}$$

$$\Delta P_{\text{g}} = 3.36 \times 10^{-6} \times 0.009 \times 77.6 \times \frac{132,277.2^2}{\frac{1}{0.002} \times (6 - 0.28)^5}$$

$$\boldsymbol{\Delta P_{\text{g}} = 0.14\text{psig} = 0.94\text{kpag}}$$

38 Economizer heat duty prediction

Economizers are heat exchangers that increase temperature of fluids, normally water. Since economizers are recovering more enthalpy from waste gas, they are improving the boiler's efficiency. They save energy from exhaust gases to preheat the cold water which will be used as the feed water [27].

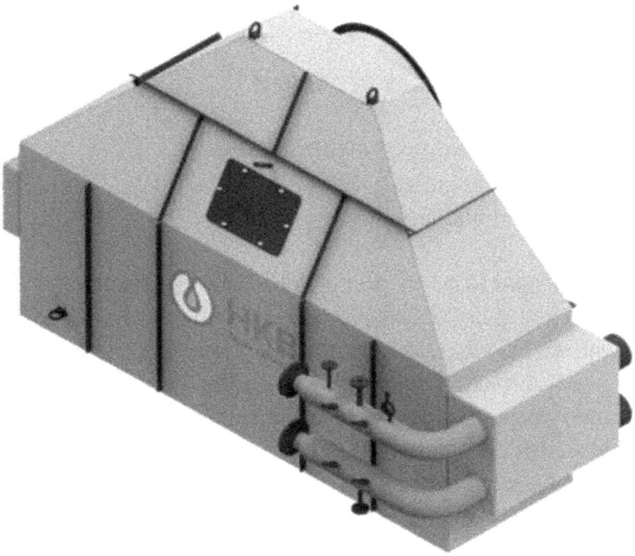

Fig. 38.1: Economizer by HKB Boiler Solution.

Economizer water inlet temperature	230	°F
Economizer inlet enthalpy	199.8	Btu/lb
Economizer water outlet temperature	345.2	°F
Economizer outlet enthalpy	317.73	Btu/lb
Blow down rate	3%	%
Economizer mass flow	136,245.516	lb/h

$$Q_{\text{duty}} = (h_{\text{outlet water}} - h_{\text{inlet water}}) \times \text{Evaporation@ } 100\%\text{MCR}$$

$$Q_{\text{duty}} = (317.73 - 199.8)\ \text{Btu/lb} \times 136{,}245.516 \times (1 + 3\%)\ \text{lb/h}$$

$$\mathbf{Q_{duty} = 16.06\ MM\ Btu/h}$$

https://doi.org/10.1515/9783110757088-038

39 Economizer tube rows deep no.

Economizer tube rows deep number will be calculating by below sequence:
- Required feed water pump outlet pressure
- Economizer design pressure
- Economizer tube thickness
- Economizer tube area
- Economizer tubes row deep number

39.1 Required feed water pump outlet pressure

Basically, there are two options for connecting economizer to boiler.
1 By putting isolating valve before economizer
2 By putting isolating valve between boiler and economizer

In option #1, economizer is connected to boiler and economizer operating pressure equal to boiler and economizer should be located at higher elevation than the boiler to not permit water coming back from steam drum. In this option, if economizer requires maintenance, boiler need to be shut down. But in option #2, economizer is separated from boiler and can be located in lower elevation from steam drum which lead to lower cost for economizer structure and foundation.

BFW (boiler feed water) pump outlet pressure can be obtaining from boiler design pressure plus chevron and economizer and interconnection piping pressure drop.

At this step, it is better to assume economizer pressure drop which after economizer calculation can be used from calculated pressure drop again.

Safety valve higher Setting	48.17	kg/cm^2
Chevron pressure drop (Maximum)	0.51	kg/cm^2
Piping pressure drop	2.10	kg/cm^2
Economizer pressure drop	0.01	kg/cm^2

$$P_{\text{pump outlet}} = P_{\text{PSV}_2} + \Delta P_{\text{chevron}} + \Delta P_{\text{piping}} + \Delta P_{\text{economizer}}$$

$$P_{\text{pump outlet}} = 48.17 + 0.51 + 2.1 + 0.01$$

$$\mathbf{P_{pump\ outlet} = 50.78\ kg/cm^2 = 722.30\ psig}$$

https://doi.org/10.1515/9783110757088-039

39.2 Economizer design pressure

The design pressure will be defined according to the below criteria except in cases approved by the company [24].

- For maximum normal operating pressure less than 1.5 barg, use 3.5 bar gage.
- For maximum normal operating pressures between 1.5 and 20 barg, use the maximum normal operating gage pressure + 2 bar.
- For maximum normal operating pressures between 20 and 80 barg, use 110% of the maximum normal operating gage pressure.

Economizer design pressure: 722.3*1.1 = 794.53 psig

39.3 Economizer tube thickness

$$t_{\min} = \frac{PD}{2S + P} + 0.005D + e$$

where t is minimum required tube thickness, inch; P is economizer design pressure, psig; D is tube diameter, inch (assume 2 in); S is maximum allowable stress for A178-Gr.A, psi; e is thickness factor for expanded tube ends as per ASME 1, PG-27.4

$$t_{\min} = \frac{794.53 \times 2}{2 \times 12,400 + 794.53} + 0.005 \times 2 + 0$$

$$t_{\min} = 0.072 \text{ in}$$

As per tube manufacture catalog for 2-in diameter, they produce tube with thickness 0.085 in. So,

$$\boldsymbol{t_{\text{selected}} = 0.085 \text{ in}}.$$

39.4 Economizer tube area

$$A_{\text{economizer tube}} = \frac{\pi}{144} \times \left(\frac{D - t}{2}\right)^2$$

$$A_{\text{economizer tube}} = \frac{\pi}{144} \times \left(\frac{2 - 0.085}{2}\right)^2$$

$$\boldsymbol{A_{\text{economizer tube}} = 0.0199 \text{ft}^2}$$

39.5 Economizer tubes row deep number

Water velocity range inside economizer tubes, 150–300 fpm [41]
Selected water velocity = 150 fpm
Water specific volume = 0.016 ft^3/lb
Water volumetric flow rate = 2,182.44 ft^3/h
Selected tube row deep no. = 12

Fig. 39.1: 75 T/h, 40 barg, 410 °C water tube boiler, Kermanshah Petrochemical, Kermanshah, Iran.

40 Economizer tube arrangement

Economizers are mostly designed in the form of bare tube, inline and crossflow arrangements. When we use coal as fuel, fly ash produces high fouling and erosive atmosphere. By using bare tube and inline arrangement, we can minimize erosion and trapping of ash. This arrangement is easiest geometry that can be kept clean by soot blowers. Please note that this arrangement has higher weight, volume and cost [28].

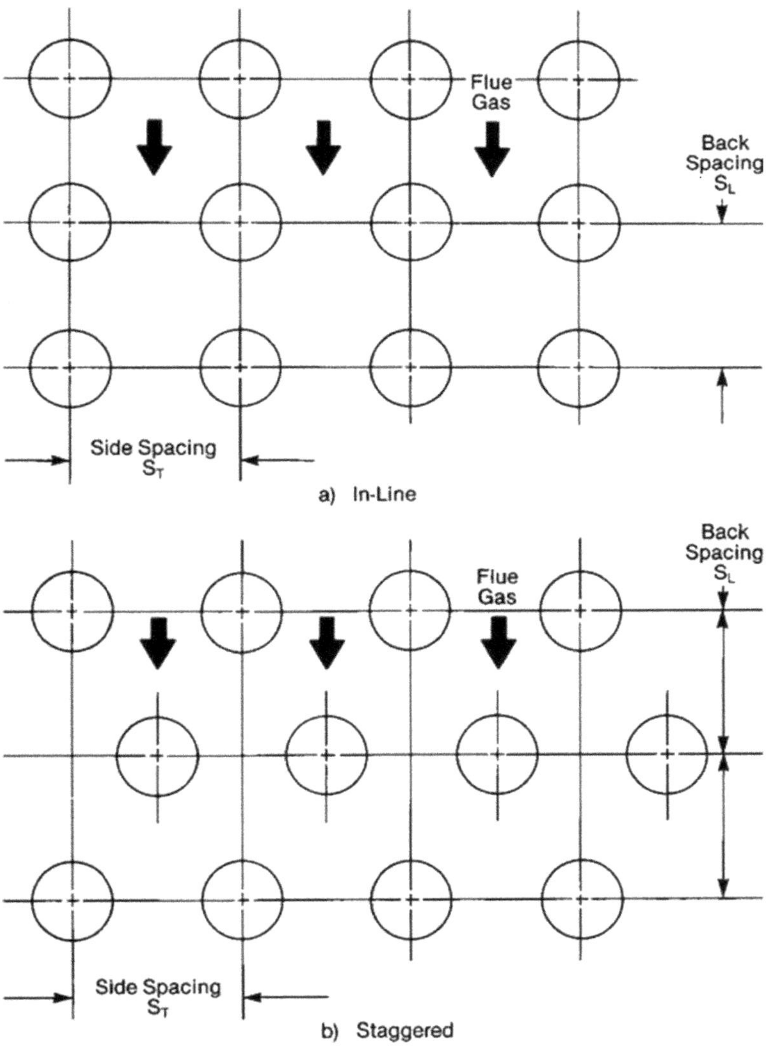

Fig. 40.1: Economizer tube arrangement [28].

https://doi.org/10.1515/9783110757088-040

41 Economizer tube solid or serrated fins

In boilers and heater fins, heat transfer surfaces are used in two types: solid and serrated. The finned tubes make heat exchangers compact. They have lower flue gas pressure drop and lower operating costs. Thermal design of an economizer should be selected according to the fin density (fins/in), height and thickness, understanding the effect on tube wall or fin temperatures and heat flux.

Fins are made of two types:

- Serrated fins, which supply higher heat transfer surface at same size, and they are cheaper. Please note that they are adequate just for using clean fuels which leave no deposits between serrations which decrease heat transfer.
- Solid fins have better cooling and can be used at final SH (superheater) and RH (reheater) sections [29, 30].

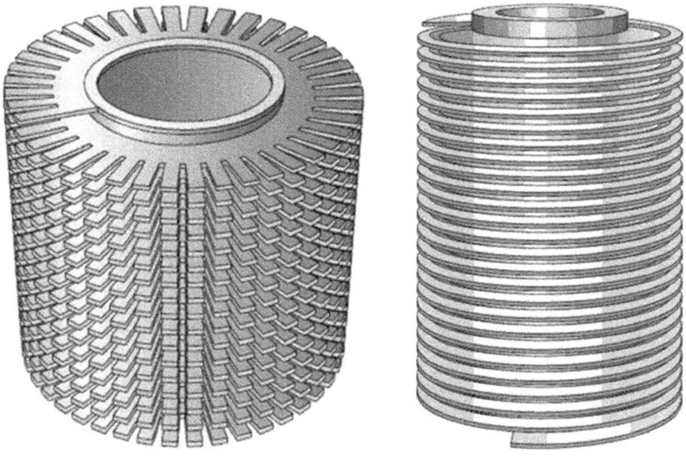

Fig. 41.1: Economizer tubes solid or serrated type.

Fin manufactures can produce carbon steel fins from characteristics provided in Tab. 41.1.

https://doi.org/10.1515/9783110757088-041

Tab. 41.1: Fin chart.

Tube OD	Fin height	Fin thickness	Fins per in
1/2	1/4	0.015	4, 5, 6, 7, 8, 9, 10
	3/8	0.015	4, 5, 6, 7, 8, 9, 10
5/8	1/4	0.015	4, 5, 6, 7, 8, 9, 10
	3/8	0.015	4, 5, 6, 7, 8, 9, 10
3/4	1/4	0.015	4, 5, 6, 7, 8, 9, 10
	3/8	0.015	4, 5, 6, 7, 8, 9, 10
	1/2	0.03	4, 5, 6, 7, 8
1	1/4	0.015	4, 5, 6, 7, 8, 9, 10
	3/8	0.015	4, 5, 6, 7, 8, 9, 10
	1/2	0.03	4, 5, 6, 7, 8
	5/8	0.03	2, 3, 4, 5, 6
1 1/4	3/8	0.015	4, 5, 6, 7, 8, 9, 10
	1/2	0.03	4, 5, 6, 7, 8
	5/8	0.03	2, 3, 4, 5, 6
1 1/2	1/2	0.03	2, 3, 4, 5, 6, 8
	5/8	0.03	2, 3, 4, 5, 6
	3/4	0.04	2, 3, 4, 5, 6
1 3/4	5/8	0.04	2, 3, 4, 5, 6
	3/4	0.04	2, 3, 4, 5, 6
2	3/4	0.04	2, 3, 4, 5, 6
	1	0.04	2, 3, 4, 5, 6

In designing process of economizer, some assumptions should be considered and performance of economizer needs to be checked. Our assumptions can be found follows:

Arrangement type (inline or staggered)	Inline	
Fin type (solid or serrated)	Solid	
Fin density	6	Fin/in
Fin height	0.75	in
Fin thickness	0.04	in
Transverse pitch	4.5	in
Longitudinal pitch	4.5	in
Tube length	10	ft

42 Economizer convection heat transfer coefficient

Economizer convection heat transfer coefficient will be calculated by the following sequence [5]:

- Economizer obstruction surface area
- Economizer gas mass velocity
- Economizer inlet flue gas properties
- Economizer flue gas Reynolds number
- Economizer convection heat transfer coefficient

42.1 Economizer obstruction surface area

$$A_{\mathrm{obs}} = \frac{d}{12} + \frac{n_{\mathrm{fin}} b H_{\mathrm{fin}}}{6}$$

$$A_{\mathrm{obs}} = \frac{2}{12} + \frac{6 \times 0.04 \times 0.75}{6}$$

$$\boldsymbol{A_{\mathrm{obs}} = 0.1967\ \mathrm{ft}^2/\mathrm{ft}}$$

42.2 Economizer gas mass velocity

$$G = \frac{W_{\text{flue gas}}}{\left(\frac{S_{\mathrm{T}}}{12} - A_{\mathrm{obs}}\right) \times N_{\text{row deep}} \times L_{\text{fin tube}}}$$

$$G = \frac{170{,}818.92}{\left(\frac{4.5}{12} - 0.1967\right) \times 12 \times 10}$$

$$\boldsymbol{G = 7{,}982.19\ \mathrm{lb}/\mathrm{ft}^2\,\mathrm{h}}$$

42.3 Economizer inlet flue gas properties

Flue gas properties at inlet temperature can be found from Tab. 42.1:

$$T_{\mathrm{gas}_{\mathrm{inlet}}} = 662.63\ ^{\circ}\mathrm{F}$$

https://doi.org/10.1515/9783110757088-042

Tab. 42.1: Flue gas product properties.

Temp (°F)	C_P	μ	K
800,00	0.29070	0.07500	0.02870
700,00	0.28695	0.07063	0.02680
600,00	0.28320	0.06625	0.02490
500,00	0.27945	0.06188	0.02300

$$\mu = 0.0706\ \text{lb/ft h}$$

$$C_p = 0.2870\ \text{Btu/lb F}$$

$$K = 0.0268\ \text{Btu/ft h F}$$

42.4 Economizer flue gas Reynolds number

$$\text{Re} = \frac{Gd_{\text{Tube}}}{12\mu}$$

$$\text{Re} = \frac{7,982.19 \times 2}{12 \times 0.0706}$$

$$\mathbf{Re = 18,837.03}$$

42.5 Economizer convection heat transfer coefficient

ESCOA Corporation uses from ESCOA correlations to evaluate heat transfer and pressure drop at arrangement of inline and staggered with solid and serrated finned tubes [5]:

$$h_c = C_3 C_1 C_5 \times \left(\frac{d + 2h}{d}\right)^{0.5} \times \left(\frac{t_{\text{flue gas}_{\text{average}}} + 460}{t_{\text{water average}} + 460}\right)^{0.25} \times GC_p \times \left(\frac{K}{\mu C_p}\right)^{0.67}$$

$$s = \frac{1}{n_{\text{fin density}}} - b_{\text{fin thickness}}$$

$$C_1 = 0.25 \times \text{Re}^{-0.35}$$

$$C_2 = 0.07 + 8\text{Re}^{-0.45}$$

42.5.1 Solid inline arrangement

$$C_3 = 0.2 + 0.65e^{-0.25\ h/s}$$

$$C_4 = 0.08 \times (0.15\, S_T/d)^{-1.1 \times (h/s)^{0.15}}$$

$$C_5 = 1.1 - \left(0.75 - 1.5e^{-0.7N_{\text{tube row depp}} \times d}\right) \times e^{-2 \times (S_L/S_T)}$$

$$C_6 = 1.6 - \left(0.75 - 1.5e^{-0.7N_{\text{tube row depp}} \times d}\right) \times e^{-2 \times (S_L/S_T)^2}$$

42.5.2 Solid staggered arrangement

$$C_3 = 0.35 + 0.65e^{-0.25\ h/s}$$

$$C_4 = 0.11 \times (0.15\ S_T/d)^{-0.7 \times (h/s)^{0.2}}$$

$$C_5 = 0.7 + \left(0.7 - 0.8e^{-0.15N_{\text{tube row depp}} \times d^2}\right) \times e^{-1 \times (S_L/S_T)}$$

$$C_6 = 1.1 + \left(1.8 - 2.1e^{-0.15N_{\text{tube row depp}} \times d^2}\right)$$
$$\times e^{-2 \times (S_L/S_T)^2} - \left(0.7 - 0.8e^{-0.15N_{\text{row deep}} \times d^2}\right) \times e^{-0.6 \times (S_L/S_T)}$$

42.5.3 Serrated inline arrangement

$$C_3 = 0.35 + 0.5e^{-0.35(h/s)}$$

$$C_4 = 0.08 \times (0.15\ S_T/d)^{-1.1 \times (h/s)^{0.2}}$$

$$C_5 = 1.1 - \left(0.75 - 1.5e^{-0.7N_{\text{tube row depp}} \times d}\right) \times e^{-2 \times (S_L/S_T)}$$

$$C_6 = 1.6 - \left(0.75 - 1.5e^{-0.7N_{\text{tube row depp}} \times d}\right) \times e^{-2 \times (S_L/S_T)^2}$$

42.5.4 Serrated staggered arrangement

$$C_3 = 0.55 + 0.45e^{-0.35(h/s)}$$

$$C_4 = 0.11 \times (0.05\, S_T/d)^{-0.7 \times (h/s)^{0.23}}$$

$$C_5 = 0.7 + \left(0.7 - 0.8e^{-0.15N_{\text{tube row depp}} \times d^2}\right) \times e^{-1 \times (S_L/S_T)}$$

$$C_6 = 1.1 + \left(1.8 - 2.1e^{-0.15N_{\text{tube row depp}} \times d^2}\right)$$

$$\times e^{-2 \times (S_L/S_T)^2} - \left(0.7 - 0.8e^{-0.15N_{\text{row deep}} \times d^2}\right) \times e^{-0.6 \times (S_L/S_T)}$$

With the above calculations, the following result can be obtained:

C_1	0.0080	s	0.1266
C_2	0.1654		
Solid inline arrangement		**Serrated inline arrangement**	
C_3	0.3479	C_3	0.4129
C_4	0.3808	C_4	0.4402
C_5	0.9985	C_5	0.9985
C_6	1.4985	C_6	1.4985
Solid staggered arrangement		**Serrated staggered arrangement**	
C_3	0.4979	C_3	0.6066
C_4	0.3256	C_4	1.0997
C_5	0.9573	C_5	0.9573
C_6	0.9596	C_6	0.9596

From the above results, convective heat transfer coefficient can be obtained further.

42.5.5 Solid inline arrangement

$$h_{c_{\text{solid inline}}} = C_3 C_1 C_5 \times \left(\frac{d + 2h}{d}\right)^{0.5} \times \left(\frac{t_{\text{flue gas average}} + 460}{t_{\text{water average}} + 460}\right)^{0.25} \times GC_p \times \left(\frac{K}{\mu C_p}\right)^{0.67}$$

$$h_{c_{\text{solid inline}}} = 0.3479 \times 0.0080 \times 0.9985 \times \left(\frac{2+2\times 0.75}{2}\right)^{0.5} \times \left(\frac{\frac{662.63+320}{2}+460}{\frac{345.2+230}{2}+460}\right)^{0.25}$$

$$\times 7,982.19 \times 0.2870 \times \left(\frac{0.0268}{0.0706\times 0.2870}\right)^{0.67}$$

$$\boldsymbol{h_{c_{\text{solid inline}}} = 10.75\ \text{Btu/ft}^2\ \text{F h}}$$

42.5.6 Solid staggered arrangement

$$h_{c_{\text{solid staggered}}} = C_3 C_1 C_5 \times \left(\frac{d+2h}{d}\right)^{0.5} \times \left(\frac{t_{\text{flue gas}_{\text{average}}}+460}{t_{\text{water}_{\text{average}}}+460}\right)^{0.25} \times GC_p \times \left(\frac{K}{\mu C_p}\right)^{0.67}$$

$$h_{c_{\text{solid staggered}}} = 0.4979 \times 0.0080 \times 0.9573 \times \left(\frac{2+2\times 0.75}{2}\right)^{0.5} \times \left(\frac{\frac{662.63+320}{2}+460}{\frac{345.2+230}{2}+460}\right)^{0.25}$$

$$\times 7,982.19 \times 0.2870 \times \left(\frac{0.0268}{0.0706\times 0.2870}\right)^{0.67}$$

$$\boldsymbol{h_{c_{\text{solid staggered}}} = 14.75\ \text{Btu/ft}^2\ \text{F h}}$$

42.5.7 Serrated inline arrangement

$$h_{c_{\text{serrated inline}}} = C_3 C_1 C_5 \times \left(\frac{d+2h}{d}\right)^{0.5} \times \left(\frac{t_{\text{flue gas average}}+460}{t_{\text{water average}}+460}\right)^{0.25} \times GC_p \times \left(\frac{K}{\mu C_p}\right)^{0.67}$$

$$h_{c_{\text{serrated inline}}} = 0.4129 \times 0.0080 \times 0.9985 \times \left(\frac{2+2\times 0.75}{2}\right)^{0.5} \times \left(\frac{\frac{662.63+320}{2}+460}{\frac{345.2+230}{2}+460}\right)^{0.25}$$

$$\times 7,982.19 \times 0.2870 \times \left(\frac{0.0268}{0.0706\times 0.2870}\right)^{0.67}$$

$$\boldsymbol{h_{c_{\text{serrated inline}}} = 12.76\ \text{Btu/ft}^2\ \text{F h}}$$

42.5.8 Serrated staggered arrangement

$$h_{c_{\text{serrated staggered}}} = C_3 C_1 C_5 \times \left(\frac{d+2h}{d}\right)^{0.5} \times \left(\frac{t_{\text{flue gas}_{\text{average}}}+460}{t_{\text{water}_{\text{average}}}+460}\right)^{0.25} \times GC_p \times \left(\frac{K}{\mu C_p}\right)^{0.67}$$

$$h_{c_{\text{serrated staggered}}} = 0.6066 \times 0.0080 \times 0.9573 \times \left(\frac{2 + 2 \times 0.75}{2}\right)^{0.5} \times \left(\frac{\frac{662.63 + 320}{2} + 460}{\frac{345.2 + 230}{2} + 460}\right)^{0.25}$$
$$\times 7,982.19 \times 0.2870 \times \left(\frac{0.0268}{0.0706 \times 0.2870}\right)^{0.67}$$

$$\boldsymbol{h_{c_{\text{serrated staggered}}} = 17.97\,\mathrm{Btu/ft^2\ F\ h}}$$

As the design here is based on the solid inline arrangement, convective heat transfer coefficient will be 10.75 $\mathrm{Btu/ft^2\ F\ h}$

43 Economizer overall heat transfer coefficient

Economizer overall heat transfer coefficient will be calculated by the following sequence:

- Economizer fin efficiency
- Economizer fin effectiveness
- Economizer overall heat transfer coefficient

43.1 Economizer fin efficiency

43.1.1 Solid fins [5]

$$A_\mathrm{f} = \pi n \times \frac{4dh + 4h^2 + 2bd + 4bh}{24}$$

$$A_\mathrm{t} = A_\mathrm{f} + \pi \frac{d(1 - nb)}{12}$$

$$E = \frac{1}{\left\{1 + 0.002292 m^2 h^2 \times \left[\frac{(d + 2h)}{d}\right]^{0.5}\right\}}$$

$$m = \left(\frac{24 h_\mathrm{o}}{Kb}\right)^{0.5}$$

43.1.2 Serrated fins [5]

$$A_\mathrm{f} = \pi dn \times \frac{2h(ws + b) + bws}{12ws}$$

$$A_\mathrm{t} = A_\mathrm{f} + \pi \frac{d(1 - nb)}{12}$$

$$E = \frac{\tanh(mh)}{mh}$$

$$m = \left[\frac{24 h_\mathrm{o}(b + ws)}{Kbws}\right]^{0.5}$$

As, here, the design is based on the solid inline arrangement, so:

https://doi.org/10.1515/9783110757088-043

$$A_f = \pi \times 6 \times \frac{4 \times 2 \times 0.75 + 4 \times 0.75^2 + 2 \times 0.04 \times 2 + 4 \times 0.04 \times 0.75}{24}$$

$$\mathbf{A_f = 6.69 ft^2/ft}$$

↓

$$A_t = A_f + \pi \frac{d(1-nb)}{12}$$

$$A_t = 3.526 + \pi \times \frac{2(1 - 3 \times 0.105)}{12}$$

$$\mathbf{A_t = 7.09 \ ft^2/ft}$$

↓

Flue gas temperature is 662.63 °F and thermal conductivity will be 26 Btu/ft F h.

↓

$$m = \left(\frac{24h_o}{Kb}\right)^{0.5}$$

$$m = \left(\frac{24 \times (10.75 + 0)}{26 \times 0.04}\right)^{0.5}$$

$$\mathbf{m = 15.77 \ 1/ft^2}$$

↓

$$E = \frac{1}{\left\{1 + 0.002292 m^2 h^2 \times \left[\frac{(d+2h)}{d}\right]^{0.5}\right\}}$$

$$E = \frac{1}{\left\{1 + 0.002292 \times 15.77^2 \times 0.75^2 \times \left[\frac{(2 + 2 \times 0.75)}{2}\right]^{0.5}\right\}}$$

$$E = 70.2\%$$

43.2 Economizer fin effectiveness

$$\eta = 1 - (1 - E)\frac{A_f}{A_t}$$

$$\eta = 1 - (1 - 0.702)\frac{6.69}{7.09}$$

$$\eta = 71.9\%$$

43.3 Economizer overall heat transfer coefficient

Overall heat transfer coefficient, U, for extended surfaces can be obtained from [5]

$$\frac{1}{U} = \frac{A_t}{h_i A_i} + \mathrm{ff}_i \frac{A_t}{A_i} + \mathrm{ff}_o \frac{A_t}{A_w} \times \frac{d}{24 K_m} \times \ln \frac{d}{d_i} + \frac{1}{\eta h_o}$$

where

A_t is surface area of finned tube, ft^2/ft;
A_i is tube inner surface area $= \pi d_i/12$, ft^2/ft;
A_w is average wall surface area $= \pi(d + d_i)/12$, ft^2/ft;
K_m is thermal conductivity of the tube wall, Btu/ft.h.F;
$d_i.d$ is tube inner and outer diameter, inch;
$f\mathrm{f}_i$ and ff_o are fouling factor inside and outside of the tube, ft^2 h F/B tu;
$h_i.h_o$ is tube side and gas side coefficient, Btu/ft^2 h F; and
η is fin effectiveness.

Equation can be simplified t:

$$\frac{1}{U} = \frac{A_t}{h_i A_i}$$

$$\downarrow$$

$$U = \eta h_o$$

$$U = 0.718 \times 10.75$$

$$\mathbf{U = 7.73\ Btu/ft^2\ h\,F}$$

44 Economizer tubes row number

Economizer tubes row number will be calculated by the following sequence:
- Economizer log mean temperature difference prediction
- Economizer heat surface
- Economizer tubes row number
- Economizer final heat surface

44.1 Economizer log mean temperature difference prediction

$$\Delta T_{\text{log}} = \frac{\left(T_{g_{\text{inlet}}} - T_{\text{outlet water}}\right) - \left(T_{g_{\text{outlet}}} - T_{\text{Inlet water}}\right)}{\text{LN}\frac{T_{g_{\text{inlet}}} - T_{\text{outlet water}}}{T_{g_{\text{outlet}}} - T_{\text{Inlet water}}}}$$

$$\Delta T_{\text{log}} = \frac{(662.63 - 345.2) - (320 - 230)}{\text{LN}\frac{662.63 - 345.2}{320 - 230}}$$

$$\boldsymbol{\Delta T_{\text{log}} = 180.43\ {}^{\circ}\text{F}}$$

44.2 Economizer heat surface prediction

$$A_{\text{economizer}} = \frac{\text{Heat duty}}{h_{\text{c}} \times \Delta T_{\text{log}}}$$

$$A_{\text{economizer}} = \frac{16.067 \times 10^6}{10.75 \times 180.43}$$

$$\boldsymbol{A_{\text{economizer}} = 8,282.79\ \text{ft}^2}$$

44.3 Economizer tubes row number

$$N_{\text{tube row}} = \text{Round up}\left(\frac{A_{\text{economizer}}}{A_{\text{t}} \times L_{\text{fin tube}} \times N_{\text{tube row deep}}}\right)$$

$$N_{\text{tube row}} = \text{Round up}\left(\frac{8,282.79}{7.09 \times 10 \times 12}\right)$$

$$\boldsymbol{N_{\text{tube row}} = 10}$$

https://doi.org/10.1515/9783110757088-044

44.4 Economizer final heat surface

$$A_{\text{economizer}} = N_{\text{tube row deep}} \times A_{\text{t}} \times L_{\text{fin tube}} \times N_{\text{tube row}}$$

↓

Tube row number should be finalized by the economizer outlet water temperature.

↓

$$A_{\text{economizer}} = 12 \times 7.09 \times 10 \times 10$$

$$\boldsymbol{A_{\text{economizer}} = 8,512.54\ \text{ft}^2}$$

45 Economizer package performance

Economizer package performance prediction will be calculated by the following sequence:

- Economizer package performance prediction
- Economizer outlet water temperature
- Economizer outlet flue gas temperature

45.1 Economizer package performance prediction

Duty and exit temperatures can be predicted by number of transfer units (NTU) method. Fundamentally, the duty Q is given by [5]

$$Q = \varepsilon C_{\text{min}} \left(T_{\text{gas inlet}} - T_{\text{steam inlet}} \right)$$

where ε depends on the type of flow, which can be counter flow or parallel flow or crossflow. In economizer, usually a counter flow arrangement is adopted. ε for this is given by

$$\varepsilon = \frac{1 - \exp[-\text{NTU} \times (1 - C)]}{1 - C \times \exp[-\text{NTU} \times (1 - C)]}$$

where

$$\text{NTU} = \frac{UA}{C_{\text{min}}}$$

$$C = \frac{C_{\text{min}}}{C_{\text{max}}}$$

$$C_{\text{min}} = \left(W' C_{\text{p}} \right)_{\text{min}}$$

$$C_{\text{max}} = \left(W' C_{\text{p}} \right)_{\text{max}}$$

At our design case,

Inlet water heat capacity @ constant volume = 0.886 Btu/lbF
Inlet flue gas heat capacity = 0.2870 Btu/lbF

$$W'_{\text{BFW}} = 136,245.52 \, \text{lb/h}$$

$$W'_{\text{flue gas}} = 170,818.2 \, \text{lb/h}$$

$$C_{\text{max}} = 136,245.52 \times 0.886 = 120,729.69$$

https://doi.org/10.1515/9783110757088-045

$$C_{min} = 0.2870 \times 170,818.2 \times \left(1 - \frac{2}{100}\right) = 48,036.16$$

↓

$$C = \frac{48,036.16}{120,729.69} = 0.4$$

Since gas temperature is low, nonluminous heat transfer is also low and can be neglected.

↓

$$\text{NTU} = \frac{7.74 \times 8,512.54}{48,036.16} = 1.37$$

↓

$$\varepsilon = \frac{1 - \exp[-1.37 \times (1 - 0.4)]}{1 - 0.4 \times \exp[-1.37 \times (1 - 0.4)]} = 0.68$$

↓

$$Q = 0.68 \times 48,036.16 \times (662.63 - 230)$$

$$\mathbf{Q = 14.14\ MM\ Btu/h}$$

45.2 Economizer outlet water temperature

$$T_{\text{water outlet}} = T_{\text{water inlet}} + \frac{Q}{C_p \times W'_{\text{water}}}$$

$$T_{\text{water outlet}} = 230 + \frac{14.14 \times 10^6}{136,245.52 \times 0.886}$$

$$\mathbf{T_{\text{water outlet}} = 347.14\,^{\circ}F}$$

So, outlet temperature is bigger and nearer to what we expect from our economizer design, which is acceptable. If not, it should change tube diameter or tube length.

45.3 Economizer outlet flue gas temperature

$$T_{\text{gas outlet}} = T_{\text{gas inlet}} - \frac{Q}{C_p \times W'_{\text{gas}}}$$

$$T_{\text{gas outlet}} = 662.63 - \frac{14.14 \times 10^6}{0.2870 \times 170,818.2}$$

$$\mathbf{T_{\text{gas outlet}} = 368.23\,^\circ F}$$

46 Economizer headers water pressure drop

Economizer headers water pressure drops will be calculated by the following sequence:
- Economizer header diameter
- Economizer header design pressure
- Economizer header thickness
- Economizer header water flow Reynolds number
- Economizer header water flow friction factor
- Economizer headers water pressure drops

46.1 Economizer header diameter

Water line velocity range inside pipes, 2.5–3.8 m/s [41]
Selected water velocity = 2.5 m/s
Water flow = 136,245.52/3.6 = 17.16 kg/s
Water specific volume = 0.001 m^3/kg
Water volumetric flow rate = 0.017 m^3/s
Area required = 0.017/2.5 = 0.0068 m^2
Required header diameter = 3.68 in
Selected header diameter = 3 in

$$\text{Nominal water velocity} = \frac{0.017}{\frac{\pi}{4} \times \left(\frac{3 \times 1{,}000}{25.4}\right)^2} = 3.76$$

Then, diameter selection is okay.

46.2 Economizer header design pressure

The design pressure will be defined according to the following criteria, except in cases approved by the company [24].
- For maximum normal operating pressure less than 1.5 barg, use 3.5 bar gage.
- For maximum normal operating pressures between 1.5 and 20 barg, use the maximum normal operating gage pressure +2 bar.
- For maximum normal operating pressures between 20 and 80 barg, use 110% of the maximum normal operating gage pressure.

Economizer operating pressure = 722.3 psig = 49.8 barg
Economizer header design pressure: 49.8 × 1.1 = 54.78 barg

https://doi.org/10.1515/9783110757088-046

46.3 Economizer header thickness

$$t_{\min} = \frac{PD}{2\times(68.947\times S\times E + P\times Y)} + A$$

where t is the minimum required header thickness, inch; P is header design pressure, barg; D is header diameter, mm; S is the maximum allowable stress for A106-Gr.B, Ksi; E is joint efficiency; Y is temperature coefficient; A is corrosion allowance, mm

$$t_{\min} = \frac{54.78\times 3\times 25.4}{2\times(68.947\times 17.1\times 0.85 + 54.78\times 0.4)} + 3$$

$$t_{\min} = 5.038\ \text{mm} = 0.198\ \text{in}$$

Minimum required thickness for calculation will be 0.198 in which can be obtained from pipe manufacturer catalog, where we can select 0.216 inch thickness or schedule 40 for 3 in diameter header.

46.4 Economizer header water flow Reynolds number

$$\text{Re} = 15.2\times\frac{W'}{\mu\times d_i}$$

$$\text{Re} = 15.2\times\frac{136{,}245.52}{0.391\times(3\times 0.216)}$$

$$\mathbf{Re = 2{,}057{,}803}$$

46.5 Economizer header water flow friction factor

The friction factor of turbulent flow of water inside header is given by

$$\text{fr} = \frac{0.316}{\text{Re}^{0.25}}$$

$$\text{fr} = \frac{0.316}{2,057,803^{0.25}}$$

$$\mathbf{fr = 0.008}$$

46.6 Economizer headers water pressure drops

$$\Delta P_g = 3.36 \times 10^{-6} \times \text{fr} \times L_e \times \frac{W'^2}{\rho \times (D-t)^5}$$

$$\Delta P_g = 3.36 \times 10^{-6} \times 0.008 \times \left(\frac{12 \times 4.5}{12} + 1.5\right) \times \frac{136{,}245.52^2}{56.75 \times (3-0.216)^5}$$

$$\Delta P_g = 0.33\,\text{psig} = 2.26\,\text{kpag}$$

47 Economizer tube bundle water pressure drops

Economizer tube bundle water pressure drops will be calculated by the following sequence:

- Economizer each row deep tubes length
- Economizer tube bundle flow Reynolds number
- Economizer tube bundle flow friction factor
- Economizer tube bundle water pressure drops

47.1 Economizer each row deep tubes length

$$L_{\text{eco. package tubes}} = N_{\text{tube row}} \times L_{\text{tube}} + (N_{\text{tube row}} - 1) \times L_{\text{elbow equivalent}}$$

$$L_{\text{elbow equivalent}} = 5.20 \text{ ft}$$

$$L_{\text{eco. package tubes}} = 10 \times 10 + (10 - 1) \times \frac{16 \times (2 - 2 \times 0.085)}{12}$$

$$\mathbf{L_{\text{eco. package tubes}} = 121.96 \text{ ft}}$$

47.2 Economizer tube bundle flow Reynolds number

$$\text{Re} = 15.2 \times \frac{W'}{\mu \times d_{\text{i}}}$$

$$\text{Re} = 15.2 \times \frac{136,245.52/12}{0.391 \times (2 - 2 \times 0.085)}$$

$$\mathbf{Re = 240,639}$$

47.3 Economizer tube bundle flow friction factor

The friction factor of turbulent flow of water inside tube is given by

$$\text{fr} = \frac{0.316}{\text{Re}^{0.25}}$$

$$\text{fr} = \frac{0.316}{240,639^{0.25}}$$

$$\mathbf{fr = 0.014}$$

https://doi.org/10.1515/9783110757088-047

47.4 Economizer tube bundle water pressure drops

$$\Delta P_{\text{tube bundle}} = 3.36 \times 10^{-6} \times \text{fr} \times L_{\text{e}} \times \frac{W'^2}{\rho \times (D-t)^5}$$

$$\Delta P_{\text{tube bundle}} = 3.36 \times 10^{-6} \times 0.014 \times 121.6 \times \frac{\left(\frac{136,245.52}{12}\right)^2}{56.75 \times (2-0.085)^5}$$

$$\Delta P_{\text{tube bundle}} = 0.52 \text{ psig} = 3.55 \text{ kpag}$$

48 Economizer package water pressure drop

$$\Delta P_{\text{economizer}} = 2 \times \Delta P_{\text{header}} + \Delta P_{\text{tube bundle}}$$

$$\Delta P_{\text{economizer}} = 2 \times 0.33 + 0.52$$

$$\mathbf{\Delta P_{economizer} = 1.17\ psig = 8.09\ kpag}$$

https://doi.org/10.1515/9783110757088-048

49 Economizer package flue gas draft pressure drops

Economizer package flue gas draft pressure drops will be calculated by the following sequence:
- Economizer package flue gas average temperature
- Economizer package flue gas density
- Economizer gas side area
- Economizer package flue gas draft pressure drops

49.1 Economizer package flue gas average temperature

$$T_{\text{average}} = \frac{368.23 + 662.63}{2}$$

$$\boldsymbol{T_{\text{average}} = 515.43\ ^{\circ}\text{F}}$$

49.2 Economizer package flue gas density

$$\rho_{\text{flue gas}} = 492 \times 29.24 \times \frac{15.47 - 0.98 \times 0.0361273}{359 \times (460 + 515.43 \times 14.7)}$$

$$\boldsymbol{\rho_{\text{flue gas}} = 0.0431\ \text{lb/cu.ft}}$$

49.3 Economizer gas side area

$$A_{\text{gas side}} = \frac{A_{\text{t}} + A_{\text{f}}}{12}$$

$$A_{\text{gas side}} = \frac{6.69 + 7.09}{12}$$

$$\boldsymbol{A_{\text{gas side}} = 1.14\ \text{ft}^2}$$

49.4 Economizer package flue gas draft pressure drops

By Grimson [5]:

$$\Delta P_{\text{flue gas}} = (f + a)\frac{G^2 N_{\text{row deep}}}{\rho \times 1.083 \times 10^9}$$

https://doi.org/10.1515/9783110757088-049

where

Staggered arrangement [5]

$$f = C_2 C_4 C_6 \times \left(\frac{d + 2h}{d}\right)^{0.5}$$

Inline arrangement [5]

$$f = C_2 C_4 C_6 \times \left(\frac{d + 2h}{d}\right)$$

$$a = \frac{1 + B^2}{4N_{\text{row deep}}} \times \frac{t_{g_2} - t_{g_1}}{460 + t_g}$$

$$B = \left(\frac{A_{\text{gas side}}}{A_{\text{obs}}}\right)^2$$

↓

$$B = \left(\frac{1.14 - 0.1967}{1.14}\right)^2$$

$$\boldsymbol{B = 0.68}$$

↓

$$a = \frac{1 + B^2}{4N_{\text{row deep}}} \times \frac{t_{g_2} - t_{g_1}}{460 + t_{g_{\text{ave.}}}}$$

$$a = \frac{1 + 0.68^2}{4 \times 12} \times \frac{374.63 - 683.54}{460 + 529.08}$$

$$\boldsymbol{a = -0.009}$$

↓

$$f = C_2 C_4 C_6 \times \left(\frac{d + 2h}{d}\right)$$

$$f = 0.1654 \times 0.3808 \times 1.4985 \times \left(\frac{2 + 2 \times 0.75}{2}\right)$$

$$\boldsymbol{f = 0.165}$$

↓

$$\Delta P_{\text{flue gas}} = (0.165 - 0.009)\frac{7,982.19^2 \times 12}{0.0431 \times 1.083 \times 10^9}$$

$$\boldsymbol{\Delta P_{\text{flue gas}} = 2.55\ \text{psig} = 64.85\text{mm WG}}$$

50 Economizer outlet duct flue gas draft pressure drops

Economizer outlet duct flue gas draft pressure drops will be calculated by the following sequence:
- Economizer outlet duct flue gas density
- Economizer outlet duct dimension
- Economizer outlet duct equivalent diameter
- Economizer outlet duct flue gas Reynolds number
- Economizer outlet duct flue gas friction factor
- Economizer outlet duct flue gas draft pressure drops

50.1 Economizer outlet duct flue gas density

$$\rho_{\text{flue gas}} = 492 \times 29.24 \times \frac{15.42 - 2.55 \times 0.0361273}{359 \times (460 + 368.23 \times 14.7)}$$

$$\boldsymbol{\rho_{\text{flue gas}} = 0.0505\ \text{lb/cu.ft}}$$

50.2 Economizer outlet duct dimension

Height and width of duct in this design will be the same as boiler exit duct dimension and will be 135.07 and 19.75 in, respectively.

50.3 Economizer outlet duct equivalent diameter

$$D_{\text{eq.}} = 1.3 \times \frac{(W \times H)^{0.625}}{(W + H)^{0.25}}$$

$$D_{\text{eq.}} = 1.3 \times \frac{(19.75 \times 135.07)^{0.625}}{(19.75 + 135.07)^{0.25}}$$

$$\boldsymbol{D_{\text{eq.}} = 51.03\ \text{in}}$$

https://doi.org/10.1515/9783110757088-050

50.4 Economizer outlet duct flue gas Reynolds number

$$\mathrm{Re} = 15.2 \times \frac{W'}{\mu \times D_{\mathrm{eq.}}}$$

↓

$$\mu = 0.0575\,\mathrm{lb/ft\,h}$$

↓

$$\mathrm{Re} = 15.2 \times \frac{170{,}818.92}{0.0575 \times 51.03}$$

$$\mathbf{Re = 884{,}910.73}$$

50.5 Economizer outlet duct flue gas friction factor

The friction factor of turbulent flow is given by

$$\mathrm{fr} = \frac{0.316}{\mathrm{Re}^{0.25}}$$

$$\mathrm{fr} = \frac{0.316}{884{,}910.73^{0.25}}$$

$$\mathbf{fr = 0.01}$$

50.6 Economizer outlet duct flue gas draft pressure drops

$$\Delta P_{\mathrm{air}} = 9.3 \times 10^{-5} \times \mathrm{f} \times W'^{2} \times \frac{L_{\mathrm{duct}}}{\rho \times d_{\mathrm{eq.}}{}^{5}}$$

$$\Delta P_{\mathrm{air}} = 9.3 \times 10^{-5} \times 0.01 \times 170{,}818.92^{2} \times \frac{10}{0.0501 \times 51.03^{5}}$$

$$\mathbf{\Delta P_{air} = 0.016\,in\ WG = 0.41\,mm\ WG}$$

51 Circulation ratio

Density difference between cooler water within the downcomer circuits and the steam–water mixture in the riser tubes develops thermal head and drives mixture through boiler tubes. This head covers losses in the boiler circuits such as

- Friction loss in downcomer tubes
- Friction and flow acceleration loss in riser tubes and connecting pipes to the drum
- Gravity loss in the evaporator and riser tubes
- Losses in the drum internals

As a rule of thumb, in the drum with higher operating pressure, lower densities exist, and then boiler has smaller circulation ratio (CR).

CR is a ratio between the mass of the steam–water mixture and generated steam. CR in natural circulation boiler calculates by trial and error or iterative calculation method, which we should first assume a CR and then calculate all the losses, and then available thermal head must be balanced with the losses. This iteration must be repeated till head and losses are balanced.

First, energy absorption, pressure and temperature of generated steam and downcomer, evaporator tube and riser tubes arrangement should be clarified.

The mixture is water with temperature under boiling point. If available head and all the losses match, our assumption is correct; otherwise, another iteration must be done [5].

CR will be calculated by the following sequence:

- Furnace heat absorption
- Drum leaving steam enthalpy
- Downcomer mass flow
- Downcomer mixture enthalpy
- Downcomer specific volume
- Height from top of water wall to bottom
- Height of water from steam drum to bottom
- Boiler water available head
- Boiling height
- Gravity loss in boiling height
- Downcomer boiling height friction loss (single phase)
- Water wall except boiling height friction loss (two phase)
- Water wall tube acceleration loss (two phase)
- Riser tube gravity loss (two phase)
- Total two-phase pressure loss
- Riser circuit heated tube friction loss
- Total losses
- CR test

https://doi.org/10.1515/9783110757088-051

CR usually at low-pressure boilers (<1,000 psia) will be in the range from 20 to 50, and high-pressure boilers (1,000–2,700 psia) will be in the range from 9 to 5. Calculation will be as follows:

$$\mathrm{CR} = \frac{1}{x}$$

flow through the evaporator = CR × the steam generated

$$\mathbf{CR_{assumed} = 40}$$

$$Q_{\text{Leaving steam from drum}} = \frac{1}{40} = 0.025$$

51.1 Furnace heat absorption

$$Q_{\text{furnace absorption}} = W_f \mathrm{LHV} - W_g h_{\text{average}}$$

$$Q_{\text{furnace absorption}} = 3,835.1 \times 2.20462 \times 19,747.8 - 170,818.92 \times 810.54$$

$$\boldsymbol{Q_{\text{furnace absorption}} = 28,511,278\ \mathbf{Btu/h}}$$

51.2 Drum leaving steam enthalpy

$$h_{\text{drum leaving steam}} = \frac{1}{\mathrm{CR}} \times h_{\text{saturated vapor}} + \left(1 - \frac{1}{\mathrm{CR}}\right) \times h_{\text{saturated liquid}}$$

$$h_{\text{drum leaving steam}} = \frac{1}{40} \times \frac{2,798.08}{2.326} + \left(1 - \frac{1}{40}\right) \times \frac{1,210.91}{2.326}$$

$$\boldsymbol{h_{\text{drum leaving steam}} = 499.93\ \mathbf{Btu/lb}}$$

51.3 Downcomer mass flow

$$m'_{\text{downcomer}} = \mathrm{CR} \times m'_{\text{BFW}}$$

$$m'_{\text{downcomer}} = \mathrm{CR} \times m'_{\text{steam}} \times (1 + \%_{\text{blowdown}})$$

$$m'_{\text{downcomer}} = 40 \times 136,245.52$$

$$\boldsymbol{m'_{\text{downcomer}} = 5,449,820.64\ \mathbf{lb/h}}$$

51.4 Downcomer mixture enthalpy

Enthalpy of entering mixture to the downcomer tubes will be calculated through an energy balance at the drum as follows:

$$h_{\text{feed water}} + \text{CR} \times h_{\text{drum leaving steam}} = h_{\text{saturated vapor}} + \text{CR} \times h_{\text{downcomer mixture}}$$

$$\frac{739.04}{2.326} + 40 \times 499.93 = \frac{2,798.08}{2.326} + 40 \times h_{\text{downcomer mixture}}$$

$$\boldsymbol{h_{\text{downcomer mixture}} = 477.8\ \text{Btu/lb}}$$

51.5 Downcomer specific volume

$$\vartheta_{\text{downcomer}} = \frac{1}{\text{CR}} \times \vartheta_{\text{saturated vapor}} + \left(1 - \frac{1}{\text{CR}}\right) \times \vartheta_{\text{saturated liquid}}$$

$$\vartheta_{\text{downcomer}} = \frac{1}{40} \times 0.0443 \times 16.0185 + \left(1 - \frac{1}{40}\right) \times 0.0013 \times 16.0185$$

$$\boldsymbol{\vartheta_{\text{downcomer}} = 0.037\ \text{cu.ft/lb}}$$

51.6 Height from top of water wall to bottom

$$H_{\text{top of water wall to bottom}} = H_{\text{furnace}} + \frac{D_{\text{mud drum}}}{2} - H_{\text{upto drilling mud drum}}$$

$$H_{\text{top of water wall to bottom}} = \frac{144 + \frac{30}{2} - 8.37}{12}$$

$$\boldsymbol{H_{\text{top of water wall to bottom}} = 12.57\ \text{ft}}$$

51.7 Height of water from steam drum to bottom

$$H_{\text{steam drum to bottom}} = H_{\text{drums CC}} + L_{\text{elbow 45}} + L_{\text{elbow 135}} + L_{\text{pipe entrance}} + L_{\text{exit}} + D_{\text{mud drum}} + D_{\text{steam drum}} - H_{\text{min allowable operating at steam drum}}$$

From later calculation

$$H_{\text{min allowable operating at steam drum}} = 14.22\,\text{in}$$

$$H_{\text{steam drum to bottom}} = \frac{163.17 + 2 \times \frac{16\times(2-0.105)}{12} + \frac{0.78\times(2-2\times0.105)}{12} + \frac{1\times(2-2\times0.105)}{12} + 30 + 48 - 8.21}{12}$$

$$\mathbf{H_{steam\ drum\ to\ bottom} = 19.83\ ft}$$

51.8 Boiler water available head

Boiler water available head can be obtained by the following formula [5]:

$$\Delta P_{\text{available head}} = \frac{H_{\text{steam drum to bottom}}}{\vartheta_{\text{downcommer}} \times 144}$$

$$\Delta P_{\text{available head}} = \frac{19.83}{0.037 \times 144}$$

$$\mathbf{\Delta P_{available\ head} = 3.67\ psig}$$

51.9 Boiling height

$$H_{\text{boiling}} = \frac{\text{CR} \times m'_{\text{generated steam}} \times H_{\text{water wall from bottom}} \times \left(h_{\text{sat.liquid}} - h_{\text{downcommer mixture}}\right)}{Q_{\text{furnace absorption}}}$$

$$H_{\text{boiling}} = \frac{40 \times 132,277.2 \times 12.57 \times \left(\frac{1,210.91}{2.326} - 477.8\right)}{28,449,783.18}$$

$$\mathbf{H_{boiling} = 9.57\ ft}$$

51.10 Gravity loss in boiling height

$$\Delta P_{\text{Gravity loss in boiling height}} = 6.95 \times 10^{-3} \times H_{\text{boiling}} \times \frac{r_4}{v_f}$$

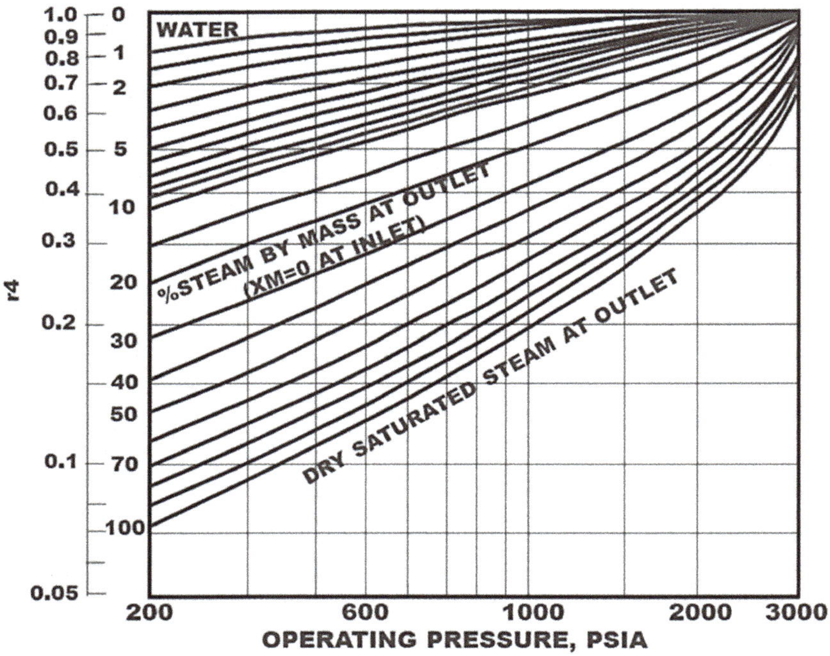

Fig. 51.1: Thom's two-phase multiplication factor for gravity loss [5].

$$r_4 = 0.7$$

$$\Delta P_{\text{Gravity loss in boiling height}} = 6.95 \times 10^{-3} \times 9.57 \times \frac{0.77}{0.037}$$

$$\Delta P_{\text{Gravity loss in boiling height}} = 1.24 \text{ psig}$$

51.11 Downcomer boiling height friction loss (single phase)

$$\Delta P_{\text{Downcomer tube boiling height}} = \frac{12 \times f \times L_{\text{water wall tube at boiling height}} \times V_{\text{boiling height}}^{2}}{2 \times g \times v_{\text{boiling phase}} \times (D - 2 \times t)}$$

$$L_{\text{water wall tube at boiling height}} = H_{\text{boiling}} + L_{45^\circ\text{elbow}} + L_{135^\circ\text{elbow}} + L_{\text{one enttrance}}$$

$$L_{\text{water wall tube at boiling height}} = H_{\text{boiling}} + \frac{16 \times (D - 2t)}{12} + \frac{16 \times (D - 2t)}{12} + \frac{K \times (D - 2t)}{12}$$

$$L_{\text{water wall tube at boiling height}} = 9.57 + 2 \times \frac{16 \times (2 - 2 \times 01.05)}{12} + \frac{0.78 \times (2 - 2 \times 01.05)}{12}$$

$$\boldsymbol{L_{\text{water wall tube at boiling height}} = 14.46\ \text{ft}}$$

$$N_{\text{rows bank tube}} = \text{Round up}\left(\frac{N_{\text{total bank tubes}}}{N_{\text{rows deep bank tube}}}\right)$$

$$N_{\text{downcomer tubes}} = \text{Round up}\left(N_{\text{rows bank tube}} \times 0.6\right) \times \left(N_{\text{rows deep bank tube}} - 1\right)$$

$$N_{\text{downcomer tubes}} = \text{Round up}(70 \times 0.6) \times (14 - 1)$$

$$\boldsymbol{N_{\text{downcomer tubes}} = 546\ \text{no.}}$$

$$V_{\text{boiling height}} = \frac{\text{CR} \times m'_{\text{generated steam}} \times v_{\text{boiling phase}}}{\frac{3.14 \times 3,600 \times N_{\text{downcommer tubes}} \times D^2}{4 \times 144}}$$

$$V_{\text{boiling height}} = \frac{40 \times 132,277.2 \times 0.037}{\frac{3.14 \times 3,600 \times 546 \times 2^2}{4 \times 144}}$$

$$\boldsymbol{V_{\text{boiling height}} = 4.63\ \text{ft/s}}$$

Tab. 51.1: Pipe friction data with flow in turbulent zone [31].

Size	½″	¾″	1″	1 ¼″	1 ½″	2″	2 ½″, 3″	4″	6″	8″–10″	12″–16″
Friction factor	0.027	0.025	0.023	0.022	0.021	0.019	0.018	0.017	0.015	0.014	0.013

$$\boldsymbol{f = 0.019}$$

$$\Delta P_{\text{Downcomer tube boiling height}} = \frac{12 \times 0.019 \times 14.46 \times 12 \times \left(\frac{4.63}{12}\right)^2}{2 \times 32.185 \times 0.037 \times (2 - 2 \times 0.105)}$$

$$\boldsymbol{\Delta P_{\text{Downcomer tube boiling height}} = 1.36 \ \text{psig}}$$

51.12 Water wall except boiling height friction loss (two phase)

$$\Delta P_{\text{Water wall tube (two phase)}} = 4 \times 10^{-10} \times v_f \times G_{\text{Water wall}}^2 \times r_3 \times \frac{f \times H_{\text{boiling}}}{D - 2 \times t}$$

$$G_{\text{Water wall}} = \frac{\text{CR} \times m'_{\text{bfw flow rate}} \times (1 + \%_{\text{blowdown}})}{\frac{3.14 \times N_{\text{downcommer tubes}} \times (D - 2 \times t)^2}{4 \times 144}}$$

$$G_{\text{Water wall}} = \frac{40 \times 136,245.52 \times (1 + 0.03)}{\frac{3.14 \times 546 \times (2 - 2 \times 0.105)^2}{4 \times 144}}$$

$$\boldsymbol{G_{\text{Water wall}} = 571.448 \ \text{lb/ft}^2 \ \text{h}}$$

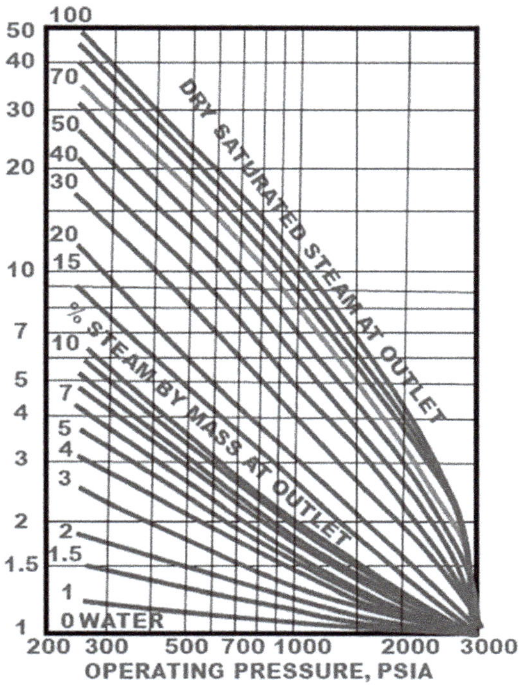

Fig. 51.2: Thom's two-phase friction factor for unheated tubes [5].

$$r_3 = 0.85$$

$$\Delta P_{\text{Waterwalltube(twophase)}} = 4 \times 10^{-10} \times 0.0013 \times 16.0185 \times 571.448^2 \times 0.85 \times \frac{0.019 \times (12.57 - 9.57)}{2 - 2 \times 0.105}$$

$$\boldsymbol{\Delta P_{\text{Water wall tube (two phase)}} = 0.07 \ \text{psig}}$$

51.13 Water wall tube acceleration loss (two phase)

$$\Delta P_{\text{Water wall tube acceleration (two phase)}} = 1.664 \times 10^{-11} \times v_f \times G_{\text{Water wall}}^2 \times r_2$$

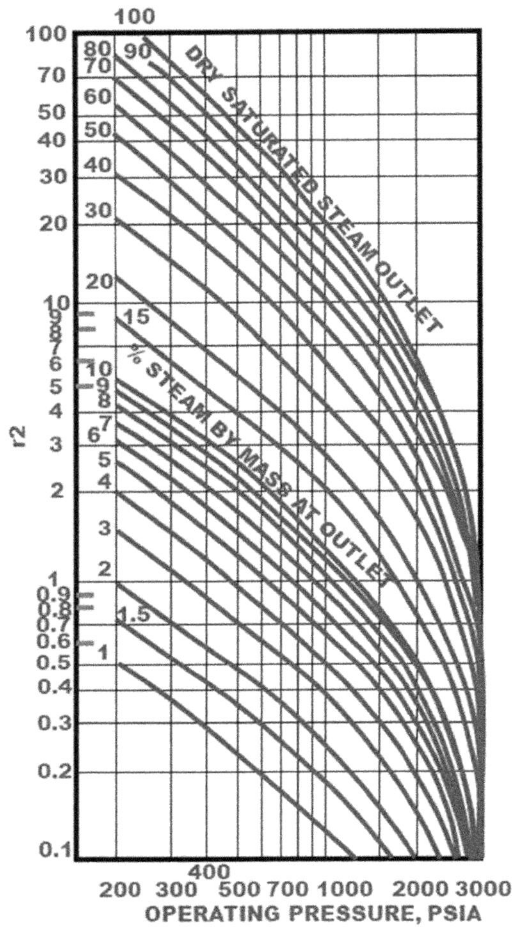

Fig. 51.3: Thom's two-phase multiplication factor for friction loss [5].

$$r_2 = 1.4$$

$$\Delta P_{\text{Water wall tube acceleration (two phase)}} = 1.664 \times 10^{-11} \times 0.0013 \times 16.0185 \times 571.448^2 \times 1.4$$

$$\boldsymbol{\Delta P_{\text{Water wall tube acceleration (two phase)}} = 0.154\ \text{psig}}$$

51.14 Riser tube gravity loss (two phase)

$$\Delta P_{\text{Riser tube gravity (two phase)}} = 6.944 \times 10^{-3} \times H_{\text{boiling}} \times \frac{r_4}{v_f}$$

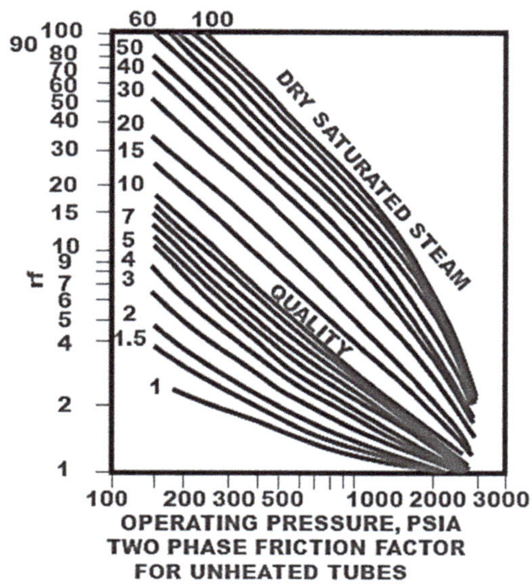

Fig. 51.4: Thom's two-phase multiplication factor for acceleration loss [5].

$$\Delta P_{\text{Riser tube gravity (two phase)}} = 6.944 \times 10^{-3} \times (12.57 - 9.57) \times \frac{0.7}{0.0013 \times 16.0185}$$

$$\mathbf{\Delta P_{\text{Riser tube gravity (two phase)}} = 0.716\ psig}$$

51.15 Total two-phase pressure loss

$$\Delta P_{\text{Total two phase}} = \Delta P_{\text{Water wall tube(two phase)}} + \Delta P_{\text{Water wall tube acceleration (two phase)}} + \Delta P_{\text{Riser tube gravity (two phase)}}$$

$$\Delta P_{\text{Total two phase}} = 0.07 + 0.154 + 0.716$$

$$\mathbf{\Delta P_{\text{Total two phase}} = 0.94\ psig}$$

51.16 Riser circuit heated tube friction loss

$$\Delta P_{\text{Riser circuit heated tube}} = f \times \frac{12 \times L_{\text{Riser circuit heated tube}}}{(D - 2 \times t)} \times G_{\text{riser}}^{2} \times \frac{v_f \times r_f}{2g \times 144}$$

$$G_{\text{riser}} = \frac{m'_{\text{steam}}}{\frac{3.14 \times N_{\text{membrane tube}} \times (D-2\times t)^2}{4\times 144}}$$

From later calculation

$$N_{\text{membrane tube}} = 110$$

$$G_{\text{riser}} = \frac{132,277.2}{\frac{3.14\times 110\times (2-2\times 0.105)^2}{4\times 144}}$$

$$\downarrow$$

$$\mathbf{G_{riser} = 68,846.05\ \ lb/ft^2\,h}$$

$$L_{\text{water wall tube at boiling height}} = L_{\text{membrane tube}} + L_{\text{one enttrance}} + L_{\text{oneexit}}$$

$$L_{\text{water wall tube at boiling height}} = L_{\text{membrane tube}} + \frac{K\times(D-2t)}{12} + \frac{K\times(D-2t)}{12}$$

From later calculation

$$L_{\text{membrane tube}} = 33.60\ \text{ft}$$

$$L_{\text{water wall tube at boiling height}} = 33.60 + \frac{0.78\times(2-2\times 0.105)}{12} + \frac{1\times(2-2\times 0.105)}{12}$$

$$\mathbf{L_{water\ wall\ tube\ at\ boiling\ height} = 33.88\ ft}$$

$$\downarrow$$

$$\mathbf{r_f = 3.5}$$

$$\downarrow$$

$$\Delta P_{\text{Riser circuit heated tube}} = 0.019 \times \frac{12\times 33.88}{(2-2\times 0.105)} \times 68,846.05^2 \times \frac{0.0013\times 16.0185\times 3.5}{2\times 32.185\times 144}$$

$$\Delta P_{\text{Riser circuit heated tube}} = 0.01\ \text{psig}$$

51.17 Total losses

$$\Delta P_{\text{total}} = \Delta P_{\text{Gravity loss in boiling height}} + \Delta P_{\text{Downcommer tube boiling height}} + \Delta P_{\text{Total two phase}} + \Delta P_{\text{Riser circuit heated tube}}$$

$$\Delta P_{\text{total}} = 1.24 + 1.36 + 0.94 + 0.01$$

$$\mathbf{\Delta P_{total} = 3.55\ psig}$$

51.18 Circulation ratio test

$$\Delta P_{\text{available head}} \cong \Delta P_{\text{total}}$$

$$\mathbf{3.66 \cong 3.55}$$

52 Flue gas stack sizing

Steel stacks are cylindrical in shape and are supported on foundation. To supply greater stability and flue gases easily, entrance must be widened in the lower portion of steel stack. The widened section decreases stresses on the steel in the base. Manufacturers design higher stack to decrease emission of flue gases. Stack height depends on the required draft.

When gases are heated, they expand and volume of gases increases. Pressure of gases will be lower ambient air pressure, then this difference results in the flow of the flue gases up the stack [32].

The steel chimneys are of two types:
- Self-supporting steel chimneys
- Guyed steel chimneys

In this book, calculations are based on self-supporting stack.

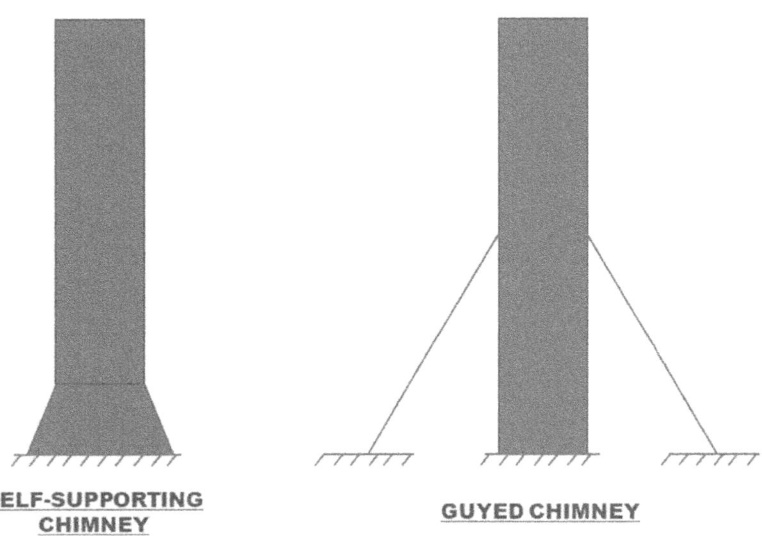

Fig. 52.1: Chimney installation type [32].

Flue gas stack sizing will be calculated by the following sequence:
- Flue gas stack diameter
- Flue gas stack height and active height

https://doi.org/10.1515/9783110757088-052

52.1 Flue gas stack diameter

Flue gas velocity ranges inside stack, 2,000–5,000 fpm [41]
Selected flue gas velocity = 2,000 fpm
Flue gas flow rate = 170,818.92 lb/h
Flue gas density = 0.0505 lb/ft^3
Flue gas volumetric flow rate = 3,385,829.51 ft^3/h
Required stack diameter = 5.995 ft = 1.827 m
Selected stack diameter = 2 m = 6.56 ft
Note: H/ID ≤ 18 to avoid excessive vibration.
Stack requested height = 20 m = 65.617 ft
H/ID = 65.617/6.56 = 10 < 18, then height and diameter is okay.

52.2 Flue gas stack height and active height

$$H_{\text{active}} = H_{\text{stack}} - \frac{\frac{D_{\text{mud drum}}}{2} + L_{\text{bank tube average}}}{12} + 1 - \frac{W_{\text{boiler exit duct}}}{2 \times 12}$$

$$H_{\text{active}} = 65.617 - \frac{\frac{30}{2} + 135.07}{12} + 1 - \frac{135.07}{2 \times 12}$$

$$\boldsymbol{H_{\text{active}} = 57.74\text{ft}}$$

Fig. 52.2: Water tube boiler (132 T/h, 42 barg, 390 °C), Basra Petrochemical, Basra, Iraq.

53 Flue gas stack net available draft

Flue gas stack net available draft will be calculated by the following sequence:
- Stack flue gas Reynolds number
- Stack flue gas friction factor
- Stack flue gas draft pressure drop
- Stack flue gas available draft
- Stack flue gas net available draft

53.1 Stack flue gas Reynolds number

$$\mathrm{Re} = 15.2 \times \frac{W'}{\mu \times D_{\mathrm{eq.}}}$$

$$\mathrm{Re} = 15.2 \times \frac{170,818.92}{0.0575 \times 6.56}$$

$$\mathbf{Re = 573,476.24}$$

53.2 Stack flue gas friction factor

The friction factor of turbulent flow is given by

$$\mathrm{fr} = \frac{0.316}{\mathrm{Re}^{0.25}}$$

$$\mathrm{fr} = \frac{0.316}{573,476.24^{0.25}}$$

$$\mathbf{fr = 0.0115}$$

53.3 Stack flue gas draft pressure drop

$$\Delta P_{\mathrm{stack}} = 9.3 \times 10^{-5} \times \mathrm{f} \times W'^2 \times \frac{L_{\mathrm{stack}}}{\rho \times d_{\mathrm{stack}}{}^5}$$

$$\Delta P_{\mathrm{stack}} = 9.3 \times 10^{-5} \times 0.0115 \times 170,818.92^2 \times \frac{65.617}{0.0505 \times (6.56 \times 12)^5}$$

$$\boldsymbol{\Delta P_{\mathrm{stack}} = 0.0134\ \mathrm{in\ WG} = 0.34\ \mathrm{mm\ WG}}$$

https://doi.org/10.1515/9783110757088-053

53.4 Stack flue gas available draft

$$\Delta P_{\text{stack draft}} = \left(\rho_{\text{ambient air}} - \rho_{\text{flue gas}}\right) \times H_{\text{active}} \times \frac{12}{64}$$

$$\Delta P_{\text{stack draft}} = (0.0752 - 0.0505) \times 57.74 \times \frac{12}{64}$$

$$\boldsymbol{\Delta P_{\text{stack draft}} = 0.275\text{in}}$$

53.5 Stack flue gas net available draft

$$\Delta P_{\text{net draft}} = \Delta P_{\text{stack draft}} - \Delta P_{\text{stack}}$$

$$\Delta P_{\text{net draft}} = 0.275 - 0.0134$$

$$\boldsymbol{\Delta P_{\text{net draft}} = 0.2616 \text{ in WG} = 6.64 \text{ mm WG}}$$

54 Stack outlet flue gas temperature

Stack outlet flue gas temperature will be calculated by the following sequence [5]:
- Stack flue gas convective heat transfer coefficient
- Stack heat surface loss
- Stack wall temperature drop across gas film
- Stack wall temperature drop across stack wall
- Stack inner wall temperature
- Stack outer wall temperature
- Stack outer wall heat transfer
- Stack flue gas temperature drop
- Barometric pressure
- Water dew point partial pressure
- Water dew point temperature
- Stack outlet flue gas temperature

54.1 Stack flue gas convective heat transfer coefficient

$$h_c = 2.44 \times \frac{W'^{0.8}}{D_{\text{stack}}{}^{1.8}} \times \left(\frac{C_P}{\mu}\right)^{0.4} \times k^{0.6}$$

$$h_c = 2.44 \times \frac{170,818.92^{0.8}}{(6.56 \times 12)^{1.8}} \times \left(\frac{0.2757}{0.0575}\right)^{0.4} \times 0.0211^{0.6}$$

$$\boldsymbol{h_c = 2.72\,\text{Btu}/\text{ft}^2\,\text{h}\,\text{F}}$$

54.2 Stack heat surface loss

$$q_{\text{stack loss}} = 0.174\varepsilon_{\text{casing}} \times \left[\left(\frac{T_{\text{casing assumed}} + 460}{100}\right)^4 - \left(\frac{T_{\text{ambient air}} + 460}{100}\right)^4\right] + 0.296 \times \left(T_{\text{casing assumed}} - T_{\text{ambient air}}\right)^{1.25} \times \left(\frac{V_{\text{wind}} + 69}{69}\right)^{0.5}$$

$$q_{\text{stack loss}} = 0.174 \times 0.9 \times \left[\left(\frac{135.2 + 460}{100}\right)^4 - \left(\frac{68 + 460}{100}\right)^4\right] + 0.296 \times (135.2 - 68)^{1.25} \times \left(\frac{33 \times 196.85 + 69}{69}\right)^{0.5}$$

$$\boldsymbol{q_{\text{stack loss}} = 630.18\,\text{Btu}/\text{ft}^2\,\text{h}}$$

https://doi.org/10.1515/9783110757088-054

54.3 Stack wall temperature drop across gas film

$$\Delta T_{\text{gas film loss}} = \frac{q_{\text{stack loss}}}{h_c}$$

$$\Delta T_{\text{gas film loss}} = \frac{630.18}{2.72}$$

$$\mathbf{\Delta T_{gas\ film\ loss} = 231.46\ F}$$

54.4 Stack wall temperature drop across stack wall

$$\Delta T_{\text{wall loss}} = q_{\text{stack loss}} \times \frac{\ln\left[\frac{D_{\text{stack}}}{\left(D_{\text{stack}} - 2 \times t_{\text{stack assumed}}\right)}\right]}{24 \times h_{\text{metal thermal}}}$$

For inlet temperature 368.23 °F, thermal conductivity will be 27.5 Btu/ft F h.

↓

$$\Delta T_{\text{wall loss}} = 630.18 \times \frac{\ln\left[\frac{78.74}{(78.74 - 2 \times 0.79)}\right]}{24 \times 27.5}$$

$$\mathbf{\Delta T_{wall\ loss} = 1.52\ ^\circ F}$$

54.5 Stack inner wall temperature

$$T_{\text{stack innerwall}} = T_{\text{gas}_{\text{inlet}}} - \Delta T_{\text{gas film loss}}$$

$$T_{\text{stack inner wall}} = 368.23 - 231.46$$

$$\mathbf{T_{stack\ inner\ wall} = 136.78\ ^\circ F}$$

54.6 Stack outer wall temperature

$$T_{\text{stack outer wall}} = T_{\text{gas}_{\text{inlet}}} - T_{\text{stack inner wall}} - \Delta T_{\text{wall loss}}$$

$$T_{\text{stack outer wall}} = 368.23 - 231.46 - 1.52$$

$$\mathbf{T_{stack\ outer\ wall} = 135.26\ ^\circ F}$$

As shown, the calculated outer wall temperature is equal to the assumed temperature. So, our guess was correct:

$$T_{\text{stack outer wall}} \cong T_{\text{assumed stack outer wall}}$$

$$\mathbf{135.26 \cong 135.2}$$

54.7 Stack outer wall heat transfer

$$Q_{\text{stack outer wall}} = q_{\text{stack loss}} \times \frac{3.4 \times D_{\text{stack}} \times H_{\text{stack active}}}{12}$$

$$Q_{\text{stack outer wall}} = 630.18 \times \frac{3.4 \times 78.74 \times 57.74}{12}$$

$$\mathbf{Q_{\text{stack outer wall}} = 749,687.46\ Btu/h}$$

54.8 Stack flue gas temperature drop

$$\Delta T_{\text{flue gas drop}} = \frac{Q_{\text{stack outer wall}}}{C_p \times W'_{\text{flue gas}}}$$

$$\Delta T_{\text{flue gas drop}} = \frac{749,687.46}{0.2757 \times 170,818.92}$$

$$\mathbf{\Delta T_{\text{flue gas drop}} = 15.92\ °F}$$

54.9 Barometric pressure

$$P_{\text{baromteric}} = \frac{101,325 \times (1 - 2.25577 \times 10^{-5} \times H_{\text{MASL}})^{5.25588}}{6,894.76}$$

$$P_{\text{baromteric}} = \frac{101,325 \times (1 - 2.25577 \times 10^{-5} \times 4)^{5.25588}}{6,894.76}$$

$$\mathbf{P_{\text{baromteric}} = 14.69\ psia}$$

54.10 Water dew point partial pressure

$$P_{\text{water dew point}} = \frac{m_{\text{Flue gas product H2O}} \times P_{\text{baromteric}}}{m_{\text{Total dry flue gas}}}$$

$$P_{\text{water dew point}} = \frac{2.08 \times 14.69}{17.13}$$

$$\boldsymbol{P_{\text{water dew point}} = 1.78\ \text{psia}}$$

54.11 Water dew point temperature

Water dew point temperature by looking at 1.78 psia at steam table can be found as 116.56 °F.

54.12 Stack outlet flue gas temperature

Calculated temperature for outlet flue gas must be higher than atmospheric dew point temperature.

$$T_{\text{stack outlet flue gas}} = T_{\text{gas inlet}} - \Delta T_{\text{flue gas drop}}$$

$$T_{\text{stack outlet flue gas}} = 368.23 - 15.92$$

$$\boldsymbol{T_{\text{stack outlet flue gas}} = 352.32\ ^{\circ}\text{F}}$$

55 Stack outlet flue gas velocity

$$V_{\text{flue gas}} = \frac{4 \times 273 \times 3,600 \times m'_{\text{flue gas}}}{2.20462 \times \pi \times (D_{\text{stack}} \times 0.3048)^2 \times \left(16.0185 \times \rho_{\text{Inlet flue gas}}\right) \times \left(273 + \frac{T_{\text{outlet flue gas}} - 32}{1.8}\right)}$$

$$V_{\text{flue gas}} = \frac{4 \times 273 \times 3,600 \times 170,818.92}{2.20462 \times \pi \times (6.56 \times 0.3048)^2 \times (16.0185 \times 0.0505) \times \left(273 + \frac{352.32 - 32}{1.8}\right)}$$

$$\boldsymbol{V_{\text{fluegas}} = 14\text{m/s} = 2,756.52\text{fpm}}$$

Note: Stack flue gas velocity, according to the velocity range table, should be in the range of 2,000–5,000 fpm or 10.2–25.4 m/s [41]. So, assumption and calculation are fine.

https://doi.org/10.1515/9783110757088-055

56 Stack insulation thickness

Insulation thermal conductivity can be found from insulation manufacturer's catalog and data sheet. Here, we used from local manufacturer and assumed it as 0.07 W/m °C [5]:

$$L_{\text{insulation prediction}-1} = h_{\text{insulation thermal conductivity}} \times \frac{\left(T_{\text{gas inlet}} - T_{\text{stack outer wall}}\right)}{12 \times q_{\text{stack loss}}}$$

$$L_{\text{insulation prediction}-1} = 0.07 \times \frac{(368.23 - 135.26)}{12 \times 630.18}$$

$$\mathbf{L_{\text{insulation prediction}-1} = 0.29\ ft = 3.54\ in}$$

↓

$$L_{\text{insulation prediction}-2} = \frac{D_{\text{stack}} + 2 \times L_{\text{insulation assumption}}}{2 \times 12 \times \ln\left(\frac{D_{\text{stack}} + 2 \times L_{\text{insulation assumption}}}{D_{\text{stack}}}\right)}$$

$$L_{\text{insulation prediction}-2} = \frac{78.74 + 2 \times 3.35}{2 \times 12 \times \ln\left(\frac{78.74 + 2 \times 3.35}{78.74}\right)}$$

$$L_{\text{insulation prediction}-2} = 0.29\ \text{ft}$$

$$L_{\text{insulation prediction}-1} = L_{\text{insulation prediction}-2}$$

↓

$$\mathbf{L_{\text{insulation}} = 0.29\ ft = 3.35\ in = 85.09\ mm}$$

https://doi.org/10.1515/9783110757088-056

57 Force draft fan electric driver

Forced draft fan electric driver selection will be calculated by the following sequence:
- System gas pressure loss calculation
- Forced draft fan test block condition
- Forced draft fan brake horse power
- Forced draft fan required horse power
- Forced draft fan electric driver selection

57.1 System gas pressure loss calculation

By collecting all draft pressure drops at all sections, total draft pressure loss is given in Tab. 57.1.

Tab. 57.1: System pressure loss.

Inlet screen	0.00	in WG
Silencer	0.00	in WG
Fd. fan discharge duct draft loss	0.19	in WG
Damper	0.00	in WG
Air measurement	0.00	in WG
Air heater	0.00	in WG
Air preheater	0.00	in WG
Burner, windbox draft loss	9.84	in WG
Furnace draft pressure drop	0.01	in WG
Superheater draft loss	1.40	in WG
Bank tube draft loss	1.57	in WG
Boiler bank exit duct draft loss	0.08	in WG
Economizer draft loss	2.55	in WG
Economizer exit duct draft loss	0.02	in WG
Stack	0.01	in WG
Miscellaneous	0.50	in WG
Total	16.18	**in WG**

https://doi.org/10.1515/9783110757088-057

57.2 Forced draft fan test block condition

The fan should be capable of supplying the following extra capacity:

a) MCR	14% over the design air flow
	30% over the design fan head
b) PEAK	10% over the designed air flow
	21% over the designed fan head
	25 °F add to ambient temperature

Net condition

$$W'_{\mathrm{a}} = 162,683.59\ (\mathrm{lb/h}) = 73,792.12\ (\mathrm{kg/h})$$

$$V_{\mathrm{a}} = 36,048.2\ (\mathrm{SCFM})$$

$$P_{\mathrm{static}} = 16.18\ \mathrm{in\ WG}$$

$$T_{\mathrm{ambient}} = 68\ ^{\circ}\mathrm{F}$$

$$\rho_{\mathrm{air}} = 0.0752\ \mathrm{lb/cu.ft}$$

↓

Test block condition

$$W'_{\mathrm{a}} = 162,683.59(\mathrm{lb/h}) = 73,792.12(\mathrm{kg/h})$$

$$V_{\mathrm{a}} = 39,653.02(\mathrm{SCFM})$$

$$P_{\mathrm{static}} = 19.58\ \mathrm{in\ WG}$$

$$T_{\mathrm{ambient}} = 93\ ^{\circ}\mathrm{F}$$

$$\rho_{\mathrm{air}} = 0.0751\ \mathrm{lb/cu.ft}$$

57.3 Forced draft fan brake horse power

The required power to drive motor of the fan name as brake horsepower and it is a function of fan efficiency and mechanical horsepower [33]:

$$BHP = \frac{V_a \times P_{static}}{6,356 \times 100 \times \eta_{fan}}$$

$$BHP = \frac{39,653.02 \times 19.58}{6,356 \times 100 \times 0.75}$$

$$\mathbf{BHP = 174.48\ hp}$$

57.4 Forced draft fan required horse power

$$HP_{elec.} = \frac{BHP_{fan}}{\eta_{electro\ motor}}$$

$$HP_{elec.} = \frac{174.48}{0.85}$$

$$\mathbf{HP_{elec.} = 205.27\ hp = 153.44\ kW}$$

57.5 Forced draft fan electric driver selection

Fan electrodrive can be selected as higher power after required electromotor power at manufacturer chart as given in Tab. 57.2 [34].

Tab. 57.2: Electric motor chart [34].

4 Pole - 1500 rpm - 50 Hz

Part No.	Output kW	IEC Frame	Rated speed (rpm)	Full load current I_r (A)	Locked rotor current I_l/I_r	Full load torque T_r (Nm)	Locked rotor torque T_l/T_r	Break-down torque T_b/T_r	415 V % of full load Efficiency η 50	75	100	Power factor (Cos φ) 50	75	100	Sound pressure level dB (A)	Moment of Inertia J (kgm²)	Max. locked rotor time(s) Cold	Hot	Approx Weight (kg)
HGF02000433	200	315C/D/E	1485	43.9	6.3	1285	1.4	2.6	93.7	94.4	94.8	0.70	0.80	0.84	75	5.1	33	15	1780
HGF02200433	220	315C/D/E	1484	47.8	6.3	1413	1.4	2.5	94.0	94.6	94.8	0.73	0.82	0.85	75	5.4	33	15	1810
HGF02500433	250	315C/D/E	1484	54.8	6.3	1609	1.4	2.5	94.2	95.0	95.1	0.71	0.80	0.84	75	5.7	33	15	1840
HGF02800433	280	315C/D/E	1485	61.3	6.3	1805	1.4	2.6	94.3	95.2	95.2	0.71	0.81	0.84	75	6.6	33	15	1930
HGF03150433	315	315C/D/E	1486	68.8	6.8	2021	1.4	2.7	94.5	95.2	95.3	0.71	0.80	0.84	75	7.7	33	15	2050
HGF03550433	355	355C/D/E	1487	75.6	6.0	2286	1.5	2.5	94.5	95.4	95.5	0.75	0.83	0.86	79	10.7	44	20	2740
HGF04000433	400	355C/D/E	1487	84.9	6.0	2570	1.5	2.5	95.0	95.7	95.8	0.75	0.83	0.86	79	11.6	44	20	2830
HGF04500433	450	355C/D/E	1487	95.5	6.2	2894	1.6	2.5	95.1	95.8	95.9	0.74	0.82	0.86	79	13.4	44	20	2950
HGF05000433	500	400L/A/B	1490	108	6.6	3208	1.5	2.7	95.6	96.2	96.2	0.72	0.81	0.84	79	18.1	44	20	3460

The nearest power to our power is 200 kW.

58 Pressure safety valve sizing

The ASME code for boilers and pressure vessels (Sections 1 and 8) explain rules for sizing of safety or relief valves. Boilers with 500 ft^2 or more heating surface must be provided two or more safety valves. Superheaters must have minimum one safety valve and it must relieve at least 20% boiler capacity. Each steam drum's safety valve should relieve at least 75% of boiler capacity.

When two valves on steam drum installed capacity of the smaller one must be minimum 50% of the larger one, lowest valve setting pressure is at least 5% above the drum pressure. Please note that set pressure must not be more than boiler design pressure. The difference between the lowest and highest setting of boiler safety valve should not be more than 10% of the set pressure of highest setting. When each valve blows steam, that valve will be close to 97% of its setting. The highest setting cannot be set more than 3% over design pressure [35].

Pressure safety valve sizing will be calculating as per below sequence:

- Superheater safety valve sizing
- Boiler safety valve sizing

58.1 Superheater safety valve sizing

Minimum required flow for superheater safety valve can be obtaining from the below formula which is based on experiences

$$m'_{\text{required super heater PSV}} = 0.2 \times 0.18 \times m'_{\text{boiler output}}$$

$$m'_{\text{required super heater PSV}} = 0.2 \times 1.18 \times 60,000$$

$$\boldsymbol{m'_{\textbf{required super heater PSV}} = 14,160\ \textbf{kg/h} = 31,217.42\ \textbf{lb/h}}$$

$$P_{\text{super heater set}} = 1.03 \times 1.05 \times P_{\text{boiler output}}$$

$$P_{\text{super heater set}} = 1.03 \times 1.05 \times 42.82 \times 0.980665$$

$$\boldsymbol{P_{\textbf{super heater set}} = 45.42\ \textbf{barg} = 658.8\ \textbf{psig}}$$

At this step, it has to use from one PSV manufacturer, which we used from Kunkle company. PSV catalog is given in Tab. 58.1 [36].

https://doi.org/10.1515/9783110757088-058

Tab. 58.1: Safety valve sizing chart [36].

Kunkle Safety and Relief Products
Models 300 and 600

ASME Section I Steam - Capacities (U.S., lb/hr) - Flow Coefficient = 0.878

Set Pressure (psig)	Orifice Area, in^2									
	F (0.307)	G (0.503)	H (0.785)	J (1.287)	K (1.839)	L (2.853)	M (3.597)	N (4.340)	P (6.380)	Q (11.045)
15	440	720	1124	1843	2633	4085	5151	6215	9136	15816
20	509	834	1301	2134	3049	4730	5963	7195	10577	18311
25	578	948	1479	2424	3464	5374	6776	8175	12018	20806
30	648	1061	1656	2715	3880	6019	7588	9156	13459	23300
35	717	1175	1833	3006	4295	6663	8401	10136	14900	25795
40	786	1288	2011	3296	4710	7307	9213	11116	16341	28290
45	856	1402	2188	3587	5126	7952	10025	12096	17782	30784
50	925	1516	2365	3878	5541	8596	10838	13077	19223	33279
55	994	1629	2543	4168	5956	9241	11650	14057	20664	35774
60	1064	1743	2720	4459	6372	9885	12463	15037	22105	38268
65	1133	1856	2897	4750	6787	10529	13275	16017	23546	40763
70	1204	1972	3078	5046	7211	11187	14104	17017	25016	43308
75	1275	2089	3261	5346	7639	11850	14941	18027	26500	45877
80	1347	2206	3443	5645	8066	12514	15778	19037	27985	48447
85	1418	2323	3626	5945	8494	13178	16614	20046	29469	51016
90	1489	2440	3808	6244	8922	13842	17451	21056	30953	53586
95	1561	2557	3991	6543	9350	14505	18288	22066	32437	56155
100	1632	2674	4174	6843	9778	15169	19125	23075	33922	58725
125	1989	3259	5087	8340	11917	18488	23309	28124	41343	71572
150	2346	3845	6000	9837	14056	21806	27493	33172	48764	84420
175	2704	4430	6913	11334	16195	25125	31677	38220	56185	97268
200	3061	5015	7826	12831	18334	28444	35861	43269	63607	110115
225	3418	5600	8739	14328	20473	31762	40045	48317	71028	122963
250	3775	6185	9652	15825	22613	35081	44229	53365	78449	135811
275	4132	6770	10566	17322	24752	38399	48413	58413	85870	148658
300	4489	7355	11479	18819	26891	41718	52597	63462	93292	161506
325	4846	7940	12392	20316	29030	45037	56781	68510	100713	174353
350	5203	8525	13305	21813	31169	48356	60965	73558	108134	187201
375	5560	9110	14218	23310	33308	51674	65149	78607	115556	200049
400	5918	9696	15131	24807	35447	54993	69333	83655	122977	212896
425	6275	10281	16044	26304	37587	58311	73518	88703	130398	225744
450	6632	10866	16957	27801	39726	61630	77702	93752	137819	238592
475	6989	11451	17871	29299	41865	64948	81886	98800	145241	251439
500	7346	12036	18784	30796	44004	68267	86070	103848	152662	264287
525	7703	12621	19697	32293	46143	71586	90254	108897	160083	277134
550	8060	13206	20610	33790	48282	74904	94438	113945	167504	289982
575	8417	13791	21523	35287	50421	78223	98622	118993	174926	302830
600	8774	14376	22436	36784	52560	81542	102806	124042	182347	315677
625	9131	14961	23349	38281	54700	84860	106990	129090	189768	—
650	9489	15546	24262	39778	56839	88179	111174	134138	197189	—
675	9846	16132	25175	41275	58978	91498	115358	139187	204611	—
700	10203	16717	26089	42772	61117	94816	119542	144235	212032	—
725	10560	17302	27002	44269	63256	98135	123726	149283	219453	—
750	10917	17887	27915	45766	65395	101453	127910	154332	226874	—
775	11274	18472	28828	47263	67534	104772	132094	—	—	—
800	11631	19057	29741	48760	69674	108091	136278	—	—	—
825	11988	19642	30654	50257	71813	111409	140462	—	—	—
850	12345	20227	31567	51754	73952	114728	144646	—	—	—
875	12703	20812	32480	53251	76091	118047	148831	—	—	—
900	13060	21397	33394	54748	78230	121365	153015	—	—	—
925	13417	21982	34307	56245	80369	124684	157199	—	—	—
950	13774	22568	35220	57742	82508	128002	161383	—	—	—
975	14131	23153	36133	59239	84648	131321	165567	—	—	—
1000	14488	23738	37046	60737	86787	134640	169751	—	—	—

$$\boldsymbol{\dot{m}_{\text{super heater PSV}} = 39{,}778\ \text{lb/h}}$$

As per Kunkle table, orifice number *J* is obtained, and from Tab. 58.2 safety valve size will be 1½ 2½ J-orifice and weight will be 75 lb.

Tab. 58.2: Safety valve sizing chart [36].

Specifications

Model Number	Orifice[1]	Connections ANSI Standard Inlet[2]	Connections ANSI Standard Outlet[3]	Orifice Area in²	Valve Dimensions in A	Valve Dimensions in B	Valve Dimensions in C	Approximate Weight (lb)
300LFF / 600NFF	F	1 1/4"	1 1/2"	0.307	4 7/8	4 1/2	16 1/2	37
300LGF / 600NGF	G	1 1/4"	1 1/2"	0.503	4 7/8	4 1/2	16 1/2	39
300LHG / 600NHG	H	1 1/2"	2 1/2"	0.785	5 3/8	5 1/2	18 1/4	60
300LJG / 600NJG	J	1 1/2"	2 1/2"	1.287	5 3/8	5 1/2	20 7/8	75
300LKH / 600NKH	K	2"	3"	1.839	5 3/4	6	21 1/2	95
300LLJ / 600NLJ	L	2 1/2"	4"	2.853	6 5/8	6 5/8	26 1/2	133
300LMK / 600NMK	M	3"	4"	3.597	6 3/4	6 5/8	27 1/8	159
300LNM / 600NNM	N	4"	6"	4.340	8	7 1/4	33 5/8	230
300LPM / 600NPM	P	4"	6"	6.380	8	7 1/4	33 5/8	226
300LQP / 600NQP	Q	6"	8"	11.045	9 3/4	9 1/4	39 3/4	400

Dimensions are for reference only

58.2 Boiler safety valve sizing

Minimum required flow for boiler safety valves can be obtained from the following formula:

$$\dot{m}_{\text{required boiler PSV}} = 0.4 \times \dot{m}_{\text{boiler output}}$$

$$\dot{m}_{\text{required boiler PSV}} = 0.4 \times 60{,}000$$

$$\boldsymbol{\dot{m}_{\text{required boiler PSV}} = 24{,}000\ \text{kg/h} = 52{,}910.88\ \text{lb/h}}$$

$$P_{\text{boiler no. \#1set}} = 1.03 \times 1.05 \times P_{\text{drum operating}}$$

$$P_{\text{boiler no. \#1set}} = 1.03 \times 1.05 \times 43.36 \times 0.980665$$

$$\boldsymbol{P_{\text{boiler no. \#1set}} = 46.9\ \text{barg} = 680.2\ \text{psig}}$$

$$P_{\text{boilerno. \#2set}} = 1.03 \times 1.05 \times P_{\text{drum operating}}$$

$$P_{\text{boiler no. \#2set}} = (1.03 \times 1.05 \times 43.36 + 0.34) \times 0.980665$$

$$\boldsymbol{P_{\textbf{boiler no.\#2set}} = 47.23\ \ \textbf{barg} = 685.1\,\textbf{psig}}$$

At this step, it has to use from one PSV manufacturer which we used from Kunkle company. PSV catalog is given in Tab. 58.3.

Tab. 58.3: Safety valve sizing chart [36].

Kunkle Safety and Relief Products
Models 300 and 600

ASME Section I Steam - Capacities (U.S., lb/hr) - Flow Coefficient = 0.878

Set Pressure (psig)	Orifice Area, in² F (0.307)	G (0.503)	H (0.785)	J (1.287)	K (1.839)	L (2.853)	M (3.597)	N (4.340)	P (6.380)	Q (11.045)
15	440	720	1124	1843	2633	4085	5151	6215	9136	15816
20	509	834	1301	2134	3049	4730	5963	7195	10577	18311
25	578	948	1479	2424	3464	5374	6776	8175	12018	20806
30	648	1061	1656	2715	3880	6019	7588	9156	13459	23300
35	717	1175	1833	3006	4295	6663	8401	10136	14900	25795
40	786	1288	2011	3296	4710	7307	9213	11116	16341	28290
45	856	1402	2188	3587	5126	7952	10025	12096	17782	30784
50	925	1516	2365	3878	5541	8596	10838	13077	19223	33279
55	994	1629	2543	4168	5956	9241	11650	14057	20664	35774
60	1064	1743	2720	4459	6372	9885	12463	15037	22105	38268
65	1133	1856	2897	4750	6787	10529	13275	16017	23546	40763
70	1204	1972	3078	5046	7211	11187	14104	17017	25016	43308
75	1275	2089	3261	5346	7639	11850	14941	18027	26500	45877
80	1347	2206	3443	5645	8066	12514	15778	19037	27985	48447
85	1418	2323	3626	5945	8494	13178	16614	20046	29469	51016
90	1489	2440	3808	6244	8922	13842	17451	21056	30953	53586
95	1561	2557	3991	6543	9350	14505	18288	22066	32437	56155
100	1632	2674	4174	6843	9778	15169	19125	23075	33922	58725
125	1989	3259	5087	8340	11917	18488	23309	28124	41343	71572
150	2346	3845	6000	9837	14056	21806	27493	33172	48764	84420
175	2704	4430	6913	11334	16195	25125	31677	38220	56185	97268
200	3061	5015	7826	12831	18334	28444	35861	43269	63607	110115
225	3418	5600	8739	14328	20473	31762	40045	48317	71028	122963
250	3775	6185	9652	15825	22613	35081	44229	53365	78449	135811
275	4132	6770	10566	17322	24752	38399	48413	58413	85870	148658
300	4489	7355	11479	18819	26891	41718	52597	63462	93292	161506
325	4846	7940	12392	20316	29030	45037	56781	68510	100713	174353
350	5203	8525	13305	21813	31169	48355	60965	73558	108134	187201
375	5560	9110	14218	23310	33308	51674	65149	78607	115556	200049
400	5918	9696	15131	24807	35447	54993	69333	83655	122977	212896
425	6275	10281	16044	26304	37587	58311	73518	88703	130398	225744
450	6632	10866	16957	27801	39726	61630	77702	93752	137819	238592
475	6989	11451	17871	29299	41865	64948	81886	98800	145241	251439
500	7346	12036	18784	30796	44004	68267	86070	103848	152662	264287
525	7703	12621	19697	32293	46143	71586	90254	108897	160083	277134
550	8060	13206	20610	33790	48282	74904	94438	113945	167504	289982
575	8417	13791	21523	35287	50421	78223	98622	118993	174926	302830
600	8774	14376	22436	36784	52560	81542	102806	124042	182347	315677
625	9131	14961	23349	38281	54700	84860	106990	129090	189768	—
650	9489	15546	24262	39778	56839	88179	111174	134138	197189	—
675	9846	16132	25175	41275	58978	91498	115358	139187	204611	—
700	10203	16717	26089	42772	61117	94816	119542	144235	212032	—
725	10560	17302	27002	44269	63256	98135	123726	149283	219453	—
750	10917	17887	27915	45766	65395	101453	127910	154332	226874	—
775	11274	18472	28828	47263	67534	104772	132094	—	—	—
800	11631	19057	29741	48760	69674	108091	136278	—	—	—
825	11988	19642	30654	50257	71813	111409	140462	—	—	—
850	12345	20227	31567	51754	73952	114728	144646	—	—	—
875	12703	20812	32480	53251	76091	118047	148831	—	—	—
900	13060	21397	33394	54748	78230	121365	153015	—	—	—
925	13417	21982	34307	56245	80369	124684	157199	—	—	—
950	13774	22568	35220	57742	82508	128002	161383	—	—	—
975	14131	23153	36133	59239	84648	131321	165567	—	—	—
1000	14488	23738	37046	60737	86787	134640	169751	—	—	—

$$m^{\cdot}_{\text{boiler no.\#1PSV}} = \mathbf{58,978\ lb/h}$$

$$m^{\cdot}_{\text{boiler no.\#2PSV}} = \mathbf{58,978\ lb/h}$$

As per Kunkle table, orifice number *K* is obtained, and safety valve size is given in Tab. 58.4.

Tab. 58.4: Safety valve sizing chart [36].

Parameter	Superheater	Drum 1	Drum 2	Unit
Flow (selected)	39,778	58,978	58,978	lb/h
Set pressure	685.8	680.2	685.1	psig
Kunkle PSV model	1½ × 2½ J-orifice	2 × 3 K-orifice	2 × 3 K-orifice	
Weight	75	95	95	lb

Specifications

Model Number	Orifice[1]	Connections ANSI Standard Inlet[2]	Connections ANSI Standard Outlet[3]	Orifice Area in²	Valve Dimensions in A	Valve Dimensions in B	Valve Dimensions in C	Approximate Weight (lb)
300LFF / 600NFF	F	1½"	1½"	0.307	4⅞	4½	16½	37
300LGF / 600NGF	G	1½"	1½"	0.503	4⅞	4½	16½	39
300LHG / 600NHG	H	1½"	2½"	0.785	5⅜	5½	18¼	60
300LJG / 600NJG	J	1½"	2½"	1.287	5⅜	5½	20⅞	75
300LKH / 600NKH	K	2"	3"	1.839	5¾	6	21½	95
300LLJ / 600NLJ	L	2½"	4"	2.853	6⅝	6⅝	26½	133
300LMK / 600NMK	M	3"	4"	3.597	6¾	6⅝	27⅛	159
300LNM / 600NNM	N	4"	6"	4.340	8	7¼	33⅝	230
300LPM / 600NPM	P	4"	6"	6.380	8	7¼	33⅝	226
300LQP / 600NQP	Q	6"	8"	11.045	9¾	9¼	39¾	400

Dimensions are for reference only

59 Desuperheater water

Required desuperheater water can be obtained by an energy balance:

$$m^{\cdot}_{\text{boiler outlet}} \times h_{\text{outlet steam}} + m^{\cdot}_{\text{desup water}} \times h_{\text{water}} = m^{\cdot}_{\text{outlet steam}} \times h_{\text{outlet steam}}$$

$$60,000 \times 3,309.09 + m^{\cdot}_{\text{desup water}} \times 464.74 = \left(1 + m^{\cdot}_{\text{desup water}}\right) \times 3,256.73$$

$$\boldsymbol{m^{\cdot}_{\text{desup water}} = 1,125.2\ \text{kg/h}}$$

↓

$$m^{\cdot}_{\text{outlet steam}} = m^{\cdot}_{\text{desup water}} + m^{\cdot}_{\text{boiler outlet}}$$

$$m^{\cdot}_{\text{outlet steam}} = 1,125.2 + 60,000$$

$$\boldsymbol{m^{\cdot}_{\text{outlet steam}} = 61,125.2\ \text{kg/h}}$$

https://doi.org/10.1515/9783110757088-059

60 Boiler efficiency

Efficiency of the boiler will be calculated by computation of several losses such as leaving flue gases losses, unburned fuel, radiation losses, molten ash's heat loss and so on. Reader can look for more information on ASME power test code. Efficiency of boilers can be computed by two methods [5].

Boiler efficiency will be calculated using the following sequence:

- Boiler efficiency based on input–output method
- Boiler efficiency based on heat loss method

60.1 Boiler efficiency based on input–output method

$$\eta_{\text{input - output}} = 1 - \frac{Q_{\text{Free to atmosphere}}}{Q_{\text{Produce by burner}}}$$

↓

$$Q_{\text{Free to atmosphere}} = Q_{\text{Produce by burner}} - 1.02 \times (Q_{\text{Absorption by furnace}} + Q_{\text{Absorption by super heater}} + Q_{\text{Absorption by bank tube}} + Q_{\text{Absorption by economizer}} + Q_{\text{Loss from stack}})$$

$$\mathbf{Q_{\text{Produce by burner}} = 183.66 \times 10^6\ Btu/h}$$

↓

$$Q_{\text{Absorption by furnace}} = Q_{\text{Input heat}} - Q_{\text{Free heat}} = m^{\cdot}_{\text{fuel}} \times \text{LHV} - W^{\cdot}_{\text{Flue gas}} \times h_{\text{average}}$$

$$Q_{\text{Absorption by furnace}} = 3,835.1 \times 2.20462 \times 19,747.8 - 170,818.92 \times 810.54$$

$$\mathbf{Q_{\text{Absorption by furnace}} = 28,511,278\ Btu/h}$$

↓

$$Q_{\text{Absorption by super heater}} = Q_{\text{heat duty}} = 35,773,872\ \text{Btu/h}$$

↓

$$Q_{\text{Absorption by bank tube}} = Q_{\text{flue gas}} = 57,036,325\ \text{Btu/h}$$

https://doi.org/10.1515/9783110757088-060

$$Q_{\text{Absorption by economizer}} = Q_{\text{duty}} = 14,141,758\ \text{Btu/h}$$

$$Q_{\text{Loss from stack}} = Q_{\text{stack outer wall}} = 749,687\ \text{Btu/h}$$

$$Q_{\text{Free to atmosphere}} = 183.66 \times 10^6 - 1.02 \times (28,511,278 + 35,773,872 + 57,036,325 + 14,141,758 + 749,687)$$

$$\mathbf{Q_{\text{Free to atmosphere}} = 44,726,479\ Btu/h}$$

$$\eta_{\text{input - output}} = 1 - \frac{44,726,479}{183.66 \times 10^6}$$

$$\boldsymbol{\eta_{\text{input - output}} = 75.64\%}$$

60.2 Boiler efficiency based on heat loss method

Here, calculation is based on heat losses. On heat losses method, there are two ways of efficiency stating:

a) HHV
b) LHV

Hence, efficiency becomes

$$\eta_{\text{HHV}} \times \text{HHV} = \eta_{\text{LHV}} \times \text{LHV}$$

$$\eta_{\text{HHV}} = 100 - (L_1 + L_2 + L_3 + L_4 + L_5)$$

1) Dry gas losses

$$L_1 = 24 w_{dg} \frac{t_g - t_a}{HHV}$$

where
w_{dg} is dry flue gas product, lb/lb fuel;
t_g is exit flue gas temperature, °F;
t_a is ambient temperature, °F and
HHV is HHV offuel, Btu/lb

$$L_1 = 24 \times 17.13 \times \frac{352.32 - 68}{12,144.90 \times 1.8}$$

$$\mathbf{L_1 = 5.35\%}$$

2) Loss due to combustion of hydrogen and moisture in fuel

$$L_2 = (9 \times H_2 + W) \times (1,080 + 0.46 t_g - t_a) \times \frac{100}{HHV}$$

$9 \times H_2 + W =$ Water vapor formed due to combustion of fuel

$$L_2 = 2.08 \times (1,080 + 0.46 \times 352.32 - 68) \times \frac{100}{12,114.90 \times 1.8}$$

$$\mathbf{L_2 = 11.16\%}$$

3) Loss due to moisture in air

$$L_3 = 46 M w_{da} \frac{t_g - t_a}{HHV}$$

where
M is moisture in air, lb/lb dry air and
w_{dg} is dry air flow rate, lb/lb fuel

$$L_3 = 46 \times 16.18 \times 0.009 \times \frac{352.32 - 68}{12,114.90 \times 1.8}$$

$$\mathbf{L_3 = 0.08\%}$$

4) Radiation loss

$$L_4 = 10^{0.62 - 0.42 \log Q}$$

where Q is boiler heat duty, MM Btu/h;

$$L_4 = 10^{0.62 - 0.42 \log\left(40,071.4 \times \frac{3,96832}{1,000}\right)}$$

$$\mathbf{L_4 = 0.5\%}$$

5) Unburned fuel losses and formation of CO (margin)

$$L_5 = \text{Unburned (fuel gas + fuel oil) losses} + \text{formation of CO} + \text{manufacture margin}$$

$$L_5 = 0.0\% + 0.0\% + 1\%$$

↓

$$\eta_{HHV} = 100\% - (5.35\% + 11.16\% + 0.08\% + 0.5\% + 1\%)$$

$$\boldsymbol{\eta_{HHV} = 81.92\%}$$

↓

$$\eta_{HHV} \times HHV = \eta_{LHV} \times LHV$$

$$\eta_{LHV} = \frac{81.92\% \times 12,114.90 \times 1.8}{10,971 \times 1.8}$$

$$\boldsymbol{\eta_{LHV} = 90.68\%}$$

61 Boiler package water weight

Boiler weight will be calculated by the following sequence:

- Furnace total tangent tubes length (before lance)
- Bank tube package length
- Furnace total membrane tubes length
- Bank tube front and rear wall tubes length
- Furnace front and rear wall tubes length
- Bank tube total tubes length
- Boiler tubes water weight

61.1 Furnace total tangent tubes length (before lance)

In tangent tube arrangements, small diameter tubes are located close to make an arrangement by three layers: refractory, insulation and boiler casing. This arrangement has one problem that when one tube fails, repair work will be hard. When repairing is not sufficient, then broken tube must be plugged [37].

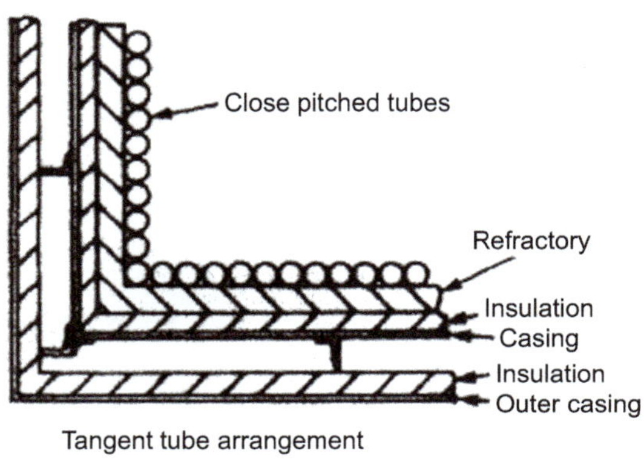

Fig. 61.1: Tangent tube arrangement [37].

$$L_{\text{furnace total tangent tubes}} = N_{\text{furnace length tangent tube}} \times L_{\text{average bank tubes}}$$

$$N_{\text{furnace length tangent tube}} = \text{round}\left(\frac{L_{\text{furnace tangent wall(before lance)}}}{D_{\text{tube}}}\right) + 1$$

https://doi.org/10.1515/9783110757088-061

$$N_{\text{furnace length tangent tube}} = \frac{400}{2} + 1$$

$$\boldsymbol{N_{\text{furnace length tangent tube}} = 201}$$

$$L_{\text{furnace total tangent tubes}} = 201 \times 135.07$$

$$\boldsymbol{L_{\text{furnace total tangent tubes}} = 27,149.37\ \text{in}}$$

61.2 Bank tube package length

$$L_{\text{bank tube}} = S_{\text{L}} \times \left(N_{\text{rows deep}\ \text{bank tube}} + 2\right)$$

$$L_{\text{bank tube}} = 4.03125 \times (70 + 2)$$

$$\boldsymbol{L_{\text{bank tube}} = 290.3\ \text{in.}}$$

61.3 Furnace total membrane tubes length

In tangent tube arrangements, small diameter tubes are located close to make an arrangement by three layers: refractory, insulation and boiler casing. This arrangement has one problem; when one tube fails, repair work will be hard. When repairing is not sufficient, then broken tube must be plugged.

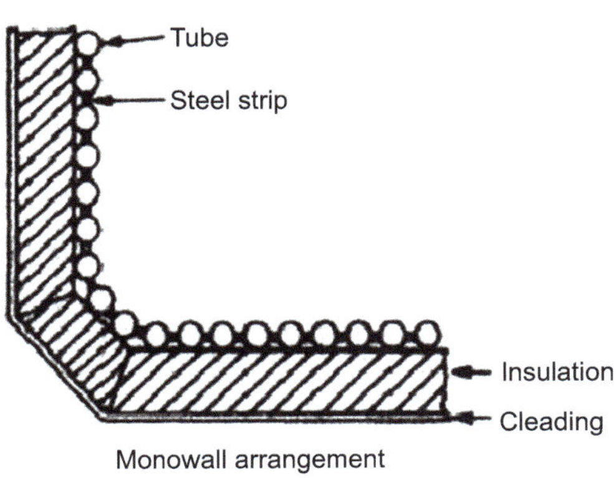

Fig. 61.2: Membrane tube arrangement [37].

In membrane wall, tubes are located together and by a steel strips weld together. This arrangement forms a gas tight structure and, on the outside, requires an insulation. With this arrangement, problems by refractory and expanded joints will be eliminated. When one tube fails, the broken tube must be plugged [37]:

$$N_{\text{furnace membrane tube}} = \frac{L_{\text{furnace}}}{S_L} = \frac{L_{\text{bank tube}} + L_{\text{super heater}} + L_{\text{return lance}}}{S_L}$$

↓

$$L_{\text{super heater}} = N_{\text{tube row}} \times S_L$$

$$L_{\text{super heater}} = 19 \times 4.03125$$

$$\boldsymbol{L_{\text{super heater}} = 73\ \text{in}}$$

↓

$$N_{\text{furnace membrane tube}} = \text{Round}\left(\frac{L_{\text{furnace}}}{S_L}\right) = \text{Round}\left(\frac{440}{4.03125}\right)$$

$$\boldsymbol{N_{\text{furnace membrane tube}} = 110}$$

↓

$$\theta_{\text{drilling tube}} = N_{\text{rows deep bank tube}} \times \theta_{\text{tube circumferential}}$$

$$\boldsymbol{\theta_{\text{drilling tube}} = 14 \times 8^\circ = 112^\circ}$$

↓

$$L_{\text{max expand to steam drum}} = \frac{D_{\text{steam drum}}}{2} \times \sin\left(\frac{180 - \theta_{\text{drilling tube}}}{2}\right)$$

$$L_{\text{max expand to steam drum}} = \frac{42}{2} \times \sin\left(\frac{180 - 112}{2}\right)$$

$$\boldsymbol{L_{\text{max expand to steam drum}} = 13.42\ \text{in}}$$

↓

$$L_{\text{max expand to mud drum}} = \frac{D_{\text{mud drum}}}{2} \times \sin\left(\frac{180 - \theta_{\text{drilling tube}}}{2}\right)$$

$$L_{\text{max expand to mud drum}} = \frac{30}{2} \times \sin\left(\frac{180 - 112}{2}\right)$$

$$\mathbf{L_{\text{max expand to mud drum}} = 8.38\ in}$$

↓

$$L_{\text{equall elbow D - type}} = 16 \times (D - 2 \times t)$$

$$L_{\text{equal elbow D - type}} = 16 \times \frac{(2 - 2 \times 0.105)}{12}$$

$$\mathbf{L_{\text{equal elbow D - type}} = 2.38\ ft}$$

↓

$$L_{\text{furnace membrane tubes}} =$$

$$H_{\text{between drums}} + \frac{2 \times L_{\text{furnace width}} + L_{\text{max expand to steam drum}} + L_{\text{max expand tomud drum}}}{12}$$

$$+ 2 \times L_{\text{equall elbow D - type}}$$

$$L_{\text{furnace membrane tubes}} = \frac{124.16 + 2 \times 100 + 13.42 + 8.38}{12} + 2 \times 2.38$$

$$\mathbf{L_{\text{furnace membrane tubes}} = 33.60\ ft = 403.26\ in}$$

↓

$$L_{\text{furnace total membrane tubes}} = N_{\text{furnace membrane tube}} \times L_{\text{furnace each membrane tube}}$$

$$L_{\text{furnace total membrane tubes}} = 110 \times 403.26$$

$$\mathbf{L_{\text{furnace total membrane tubes}} = 44,358.13\ in}$$

61.4 Bank tube front and rear wall tubes length

$$L_{\text{bank tube width}} = \frac{W_{\text{bank tube}}}{S_{\text{T}}} \times L_{\text{average bank tubes}}$$

$$L_{\text{bank tube width}} = \frac{57.26}{4.09} \times 135.07$$

$$\boldsymbol{L_{\text{bank tube width}} = 1,891\ \text{in}}$$

Fig. 61.3: Water tube boiler (132 T/h, 42 barg, 390 °C), Basra Petrochemical, Basra, Iraq.

61.5 Furnace front and rear wall tubes length

$$L_{\text{furnace width tangent tube}} = N_{\text{furnace width tangent tube}} \times H_{\text{furnace}}$$

$$N_{\text{furnace width tangent tube}} = \text{round}\left(\frac{L_{\text{furnace width}}}{D_{\text{tube}}}\right) + 1$$

$$N_{\text{furnace width tangent tube}} = \text{round}\left(\frac{100}{2}\right) + 1 = 51$$

↓

$$L_{\text{furnace width tangent tube}} = 51 \times 144$$

$$\boldsymbol{L_{\text{furnace width tangent tube}} = 7,354.2\ \text{in}}$$

Fig. 61.4: 132 T/h, 42 barg, 390 °C water tube boiler, Basra Petrochemical, Basra, Iraq.

61.6 Bank tube total tubes length

$$L_{\text{bank tube total}} = N_{\text{rows deep bank tube}} \times L_{\text{average bank tubes}} \times N_{\text{rows bank tube}}$$

$$\text{L}_{\text{bank tube total}} = 14 \times 135.07 \times 70$$

$$\boldsymbol{L_{\text{bank tube total}} = 132,370.09\ \text{in}}$$

61.7 Boiler tubes water weight

$$m_{\text{boiler tube water}} = \left(L_{\text{bank tube total}} + 1.15 \times \left(L_{\text{furnace total tangent tube}} + L_{\text{furnace total membrane tubes}} + 2 \times L_{\text{furnace Width tangent tube}} + 2 \times L_{\text{bank tube width}}\right)\right) \times \pi/4(D_{\text{tube}} - t_{\text{bank tube}})^2 \times \rho_{\text{saturated water}}$$

$$m_{\text{boiler tube water}} = (132,370.09 + 1.15 \times (27,149.37 + 44,358.13 + 2 \times 7,354.2 + 2 \times 1,891)) \times \pi/4(2 - 0.105)^2 \times \frac{788.04}{27,679.9}$$

$$\mathbf{m_{\text{boiler tube water}} = 18,929.55\ lb}$$

62 Boiler package water weight

Boiler package water weight will be calculated by the following sequence:
- Steam and mud drum length
- Mud drum water weight
- Steam drum water weight
- Boiler package water weight

62.1 Steam and mud drum length

$$L_{\text{drum}} = L_{\text{return lance}} + 2 \times L_{\text{drum head internal height}} + L_{\text{super heater}} + L_{\text{bank tube}} + 16$$

$$L_{\text{drum}} = 40 + 2 \times 13.9865 + 73 + 290.3 + 16$$

$$\boldsymbol{L_{\text{drum}} = 446.78 \text{ in}}$$

62.2 Mud drum water weight

$$m_{\text{mud drum water}} = \pi/4 \times D_{\text{mud drum}}{}^2 \times L_{\text{drum}} \times \rho_{\text{saturated water}}$$

$$m_{\text{mud drum water}} = \pi/4 \times 30^2 \times 446.78 \times \frac{788.04}{27,679.9}$$

$$\boldsymbol{m_{\text{mud drum water}} = 8,986.59 \text{ lb}}$$

62.3 Steam drum water weight

Water weight in steam drum is assumed to be 50% at the operating condition:

$$m_{\text{steam drum water}} = 1/2 \times \pi/4 \times D_{\text{steam drum}}{}^2 \times L_{\text{drum}} \times \rho_{\text{saturated water}}$$

$$m_{\text{steam drum water}} = 1/2 \times \pi/4 \times 42^2 \times 446.04 \times \frac{788.04}{27,679.9}$$

$$\boldsymbol{m_{\text{steam drum water}} = 11,502.84 \text{ lb}}$$

https://doi.org/10.1515/9783110757088-062

62.4 Boiler package water weight

$$m_{\text{boiler package water}} = m_{\text{steam drum water}} + m_{\text{mud drum water}} + m_{\text{boiler tube water}}$$

$$m_{\text{boiler package water}} = 11,502.84 + 8,986.59 + 18,929.55$$

$$\boldsymbol{m_{\text{boiler package water}} = 38,418.98 \text{ lb} = 17,880.17 \text{ kg}}$$

63 Boiler holdup time (retention time)

Boiler holdup time (retention time) will be calculated by the following sequence [5]:

- Maximum level fluctuation in steam drum according to demand
- Minimum required water up to drilling tubes inside steam drum
- Steam drum minimum diameter
- Maximum allowable level fluctuation in steam drum
- Maximum operating level fluctuation in steam drum
- Minimum level fluctuation in steam drum
- Minimum required operating water level in steam drum
- Boiler holdup time (retention time)

63.1 Maximum level fluctuation in steam drum according to demand

$$H_{\text{maximum fluctuation}} =$$

$$\frac{1{,}000 \times m'_{\text{min blowdown}} + \left(\frac{3{,}600 \times m_{\text{boiler water}} \times \left(h_{\text{sat. liquid}} - h_{\text{feed water}}\right)}{m'_{\text{min opening}} \times \left(h_{\text{sat. vapor}} - h_{\text{feed water}}\right)}\right) \times \frac{m'_{\text{max opening}} - m'_{\text{min opening}}}{3{,}600 \times t_{\text{suddenly opening}}}}{\rho_{\text{sat. liquid}} \times A_{\text{drum sectional}}}$$

$$H_{\text{maximum fluctuation}} =$$

$$1{,}000 \times \left(3\% \times 10\% \times 60{,}000 + \left(\frac{3600 \times 17{,}880.17 \times 0.238846 \times (1{,}210.91 - 739.04)}{10\% \times 60{,}000 \times 0.238846 \times (2{,}798.08 - 739.04)}\right)\right.$$
$$\left. \times \frac{60{,}000 - 10\% \times 60{,}000}{3{,}600 \times 60}\right) \times \frac{1}{788.04 \times \pi/4 \times \left(\frac{48}{0.0254}\right)^2}$$

$$\mathbf{H_{maximum\ fluctuation} = 736.68\ \ mm/s = 29\ in/s}$$

63.2 Minimum required water up to drilling tubes inside steam drum

$$H_{\text{min required up to drilling tube}} = \frac{D_{\text{internal steam drum}}}{2} \times \sin\frac{\left(180 - \theta_{\text{drilling tube}}\right)}{2}$$

https://doi.org/10.1515/9783110757088-063

$$H_{\text{min required up to drilling tube}} = 48 \times \sin\frac{(180-112)^\circ}{2}$$

$$\boldsymbol{H_{\text{min required up to drilling tube}} = 13.42\ \text{in} = 340.88\ \text{mm}}$$

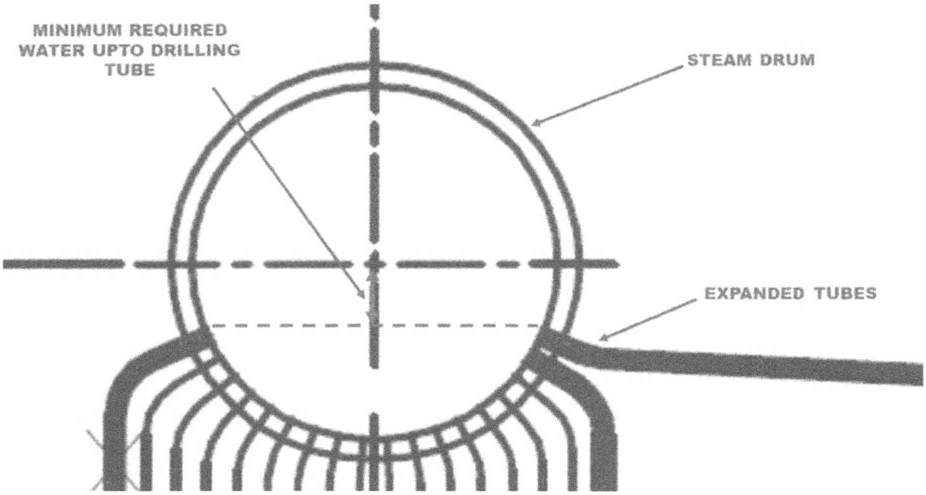

Fig. 63.1: Minimum required water up to drilling tubes.

63.3 Steam drum minimum diameter

$$D_{\text{min}} = H_{\text{min required upto drilling tube}} + H_{\text{maximum fluctuation}}$$

$$D_{\text{min}} = 13.42 + 29$$

$$\boldsymbol{D_{\text{min}} = 42.42\ \text{in}}$$

63.4 Maximum allowable level fluctuation in steam drum

$$H_{\text{maximum allowable fluctuation}} = \frac{1{,}000 \times \dfrac{3{,}600 \times m_{\text{boiler water}} \times (h_{\text{sat. liquid}} - h_{\text{feed water}})}{m'_{\text{min opening}} \times (h_{\text{sat. vapor}} - h_{\text{feed water}})} \times \dfrac{m'_{\text{max opening}} - m'_{\text{min opening}}}{3{,}600 \times t_{\text{suddenly opening}}}}{\rho_{\text{sat.liquid}} \times A_{\text{drum sectional}}}$$

$$H_{\text{maximum allowable fluctuation}} = \frac{1,000 \times \left(\left(\frac{3,600 \times 17,880.17 \times 0.238846 \times (1,210.91 - 739.04)}{10\% \times 60,000 \times 0.238846 \times (2,798.08 - 739.04)}\right) \times \frac{60,000 - 10\% \times 60,000}{3,600 \times 60}\right)}{798.70 \times \pi/4 \times \left(\frac{48}{0.0254}\right)^2}$$

$$\boldsymbol{H_{\text{maximum allowable fluctuation}} = 540.93\ \text{mm/s} = 21.29\ \text{in/s}}$$

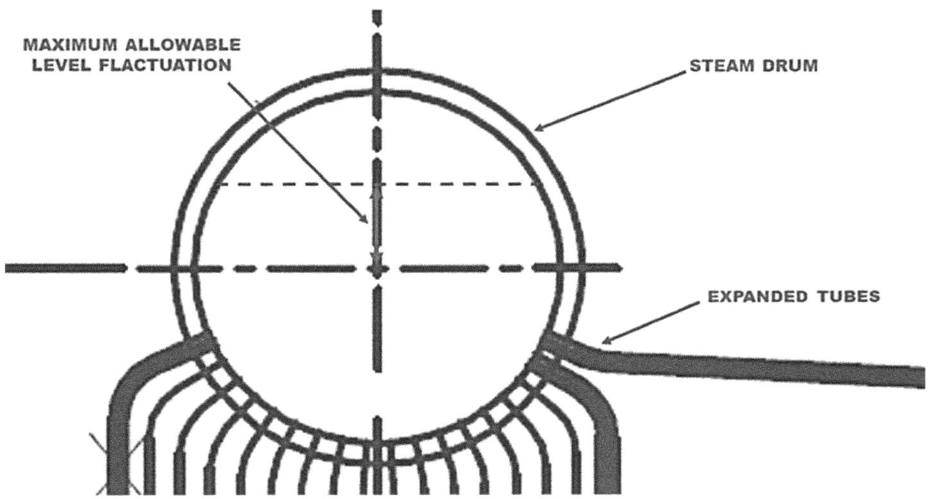

Fig. 63.2: Maximum allowable level fluctuation.

63.5 Maximum operating level fluctuation in steam drum

$$H_{\text{max allowable operating}} = H_{\text{min required upto drilling tube}} + H_{\text{maximum allowable flactuation}} - \frac{25.4}{2} \times D_{\text{steam drum}}$$

$$H_{\text{max allowable operating}} = 340.88 + 540.93 - \frac{25.4}{2} \times 48$$

$$\boldsymbol{H_{\text{max allowable operating}} = 272.21\ \text{mm} = 10.71\ \text{in upper than steam drum's center}}$$

63.6 Minimum level fluctuation in steam drum

$$H_{\text{minimum fluctuation}} = \frac{\dfrac{1,000 \times \dfrac{3,600 \times m_{\text{boiler water}} \times \left(h_{\text{sat. liquid}} - h_{\text{feed water}}\right)}{m'_{\text{min opening}} \times \left(h_{\text{sat. vapor}} - h_{\text{feed water}}\right)} \times m'_{\text{max opening}} - m'_{\text{min opening}}}{3,600 \times t_{\text{suddenly opening}}}}{\rho_{\text{sat. liquid}} \times A_{\text{drum sectional}}}$$

$$H_{\text{minimum fluctuation}} = \frac{1,000 \times \frac{3,600 \times 17,880.17 \times 0.238846 \times (1,210.91 - 739.04)}{10\% \times 60,000 \times 0.238846 \times (2,798.08 - 739.04)} \times \frac{20\% \times 60,000 - 10\% \times 60,000}{3,600 \times 60}}{788.04 \times \pi/4 \times \left(\frac{48}{0.0254}\right)^2}$$

$$\boldsymbol{H_{\text{minimum fluctuation}} = 60.1\ \text{mm/s} = 2.36\ \text{in/s}}$$

63.7 Minimum required operating water level in steam drum

$$H_{\text{min allowable operating}} = \frac{D_{\text{steam drum}}}{2} - H_{\text{min required upto drilling tube}} - H_{\text{minimum flactuation}}$$

$$H_{\text{min allowable operating}} = 25.4/2 \times 48 - 340.88 - 60.1$$

$$\boldsymbol{H_{\text{min allowable operating}} = 208.61\ \text{mm} = 8.21\ \text{in below than steam drum's center}}$$

63.8 Boiler holdup time (retention time)

$$t_{\text{hold up}} = \frac{\Delta V_{\text{drum from high tolow}}}{m'_{\text{normal operation}}}$$

↓

$$m'_{\text{normal operation}} = \frac{\vartheta_{\text{sat. liquid}} \times m'_{\text{evaporation}}}{60}$$

$$m'_{\text{normal operation}} = \frac{\frac{788.04}{27,679.9} \times 60,000 \times 2.20462}{60}$$

$$\boldsymbol{m'_{\text{normal operation}} = 62.76\ \text{ft}^3/\text{min}}$$

↓

$$\theta_{\text{up to high point}} = \text{degree}\left(\text{arc. tan}\,\frac{H_{\text{max allowable operating}}}{D/2}\right) + 90°$$

$$\theta_{\text{up to high point}} = \text{degree}\left(\text{arc. tan}\frac{10.71}{48/2}\right) + 90°$$

$$\boldsymbol{\theta_{\textbf{up to high point}} = 114.06°}$$

$$\theta_{\text{up to low point}} = \text{degree}(\text{arc. tan}\frac{H_{\text{min allowable operating}}}{D/2})$$

$$\theta_{\text{up to low point}} = \text{degree}\left(\text{arc tan}\frac{8.21}{48/2}\right)$$

$$\boldsymbol{\theta_{\textbf{up to low point}} = 18.89°}$$

$$H_{\text{between bottom and high levels}} = H_{\text{min required up to drilling tube}} + H_{\text{maximum allowable fluctuation}}$$

$$H_{\text{between bottom and high levels}} = \frac{340.88 + 540.93}{25.4}$$

$$\boldsymbol{H_{\textbf{between bottom and high levels}} = 34.72 \textbf{ in}}$$

$$V_{\text{drum head up tohigh level}} = 0.261 \times H^2{}_{\text{between bottom andhighlevels}} \times \left(3 \times \frac{D}{2} - H_{\text{between bottom and high levels}}\right)$$

$$V_{\text{drum head up to high level}} = 0.261 \times 34.72^2 \times \left(3 \times \frac{48}{2} - 34.72\right)$$

$$\boldsymbol{V_{\textbf{drum head up to high level}} = 11,728.38 \textbf{ in}^3}$$

$$\boldsymbol{H_{\textbf{between bottom andlow levels}} = H_{\textbf{min required up to drilling tube}}}$$

$$\boldsymbol{H_{\textbf{between bottom and low levels}} = 13.42 \textbf{ in}}$$

$$V_{\text{drum headup to low level}} = 0.261 \times H^2_{\text{between bottom and low levels}} \times \left(3 \times \frac{D}{2} - H_{\text{between bottom and low levels}}\right)$$

$$V_{\text{drum head up to low level}} = 0.261 \times 13.42^2 \times \left(3 \times \frac{48}{2} - 13.42\right)$$

$$\mathbf{V_{\text{drum headup to low level}} = 2,753.79\ in^3}$$

$$V_{\text{drum straight up to high level}} = L_{\text{drum}} \times \frac{D^2}{4} \times \left(\frac{\theta_{\text{up to high point}}}{57.3} - \sin\theta_{\text{up to high point}} \times \cos\theta_{\text{up to high point}}\right)$$

$$V_{\text{drum straight up to high level}} = 446.78 \times \frac{48^2}{4} \times \left(\frac{114.06}{57.3} - \sin 114.06 \times \cos 114.06\right)$$

$$\mathbf{V_{\text{drum straight up to high level}} = 391,894.48\ cu.in}$$

$$V_{\text{drum straight up to low level}} = L_{\text{drum}} \times \frac{D^2}{4} \times \left(\frac{\theta_{\text{up to low point}}}{57.3} - \sin\theta_{\text{up to low point}} \times \cos\theta_{\text{up to low point}}\right)$$

$$V_{\text{drum straightup to low level}} = 446.78 \times \frac{48^2}{4} \times \left(\frac{18.89}{57.3} - \sin 18.89 \times \cos 18.89\right)$$

$$\mathbf{V_{\text{drum straight up to low level}} = 74.049.9\ cu.in}$$

$$\Delta V_{\text{drum from high to low}} = V_{\text{drum head up to high level}} + V_{\text{drum straight up to high level}} - V_{\text{drum head up to low level}} - V_{\text{drum straight up to low level}}$$

$$\Delta V_{\text{drum from high to low}} = 11,728.38 + 391,894.48 - 2,753.79 - 74.049.9$$

$$\boldsymbol{\Delta V_{\text{drum from high to low}} = 335,793.77 \text{ cu.in} = 194.33 \text{ cu.ft}}$$

↓

$$t_{\text{holdup}} = \frac{\Delta V_{\text{drum from high to low}}}{m'_{\text{normal operation}}}$$

$$t_{\text{holdup}} = \frac{194.33}{62.76}$$

$$\boldsymbol{t_{\text{holdup}} = 3.10 \text{ min}}$$

64 Steam and mud drum weight

Steam and mud drum weight will be calculated by the following sequence:
- Steam drum elliptical heads weight
- Steam drum weight
- Mud drum elliptical heads weight
- Mud drum weight

64.1 Steam drum elliptical heads weight

$$m_{\text{steam drum heads}} = 2 \times 0.223 \times t_{\text{min}} \times (1.22 \times D_{\text{i}} + \text{SF} + t)^2$$

$$m_{\text{steam drum heads}} = 2 \times 0.223 \times 2 \times (1.22 \times 48 + 1.9865 + 2)^2$$

$$\boldsymbol{m_{\textbf{steamdrum heads}} = 3,489.56\ \textbf{lb}}$$

64.2 Steam drum weight

$$m_{\text{steam drum length}} = \pi \times D_{\text{i}} \times t \times L_{\text{steam drum}} \times \rho_{\text{steam drum}} \times (1 + m_{\text{Miscellaneous}}\%)$$

$$m_{\text{steam drum length}} = \pi \times 48 \times 2 \times 446.78 \times 0.283599 \times 1.03$$

$$\boldsymbol{m_{\textbf{steam drum length}} = 39,445.36\textbf{lb} = 17,892.1\textbf{kg}}$$

↓

$$m_{\text{steam drum}} = m_{\text{steam drum length}} + m_{\text{steam drum heads}}$$

$$m_{\text{steam drum}} = 39,445.36 + 3,489.56$$

$$\boldsymbol{m_{\textbf{steam drum}} = 42,934.92\textbf{lb} = 19,474.94\ \textbf{kg}}$$

64.3 Mud drum elliptical heads weight

$$m_{\text{mud drum heads}} = 2 \times 0.223 \times t_{\text{min}} \times (1.22 \times D_{\text{i}} + \text{SF} + t)^2$$

$$m_{\text{mud drum heads}} = 2 \times 0.223 \times 1.375 \times (1.22 \times 30 + 1.9865 + 1.375)^2$$

$$\boldsymbol{m_{\textbf{mud drum heads}} = 979.31\ \textbf{lb}}$$

https://doi.org/10.1515/9783110757088-064

64.4 Mud drum weight

$$m_{\text{mud drum length}} = \pi \times D_i \times t \times L_{\text{mud drum}} \times \rho_{\text{mud drum}} \times (1 + m_{\text{Miscellaneous}}\%)$$

$$m_{\text{mud drum length}} = \pi \times 30 \times 1.375 \times 446.78 \times 0.283599 \times 1.03$$

$$\boldsymbol{m_{\text{mud drum length}} = 16,933.57\,\text{lb} = 7,680.93\ \text{kg}}$$

↓

$$m_{\text{mud drum}} = m_{\text{mud drum length}} + m_{\text{mud drum heads}}$$

$$m_{\text{mud drum}} = 16,933.57 + 979.31$$

$$\boldsymbol{m_{\text{mud drum}} = 17,912.89\,\text{lb} = 8,125.14\ \text{kg}}$$

65 Furnace total tube number

$$N_{\text{furnace tube}} = N_{\text{tangent tube}} + N_{\text{furnace membrane tube}} + 2 \times N_{\text{furnace width tangent tube}}$$

$$N_{\text{furnace tube}} = 201 + 110 + 2 \times 51$$

$$\boldsymbol{N_{\text{furnace tube}} = 413\ \text{no.}}$$

https://doi.org/10.1515/9783110757088-065

66 Boiler total tube number

$$N_{\text{boiler tube}} = N_{\text{total bank tubes}} + N_{\text{furnace tube}} + 2 \times N_{\text{rows deep bank tube}}$$

$$N_{\text{boiler tube}} = 976 + 413 + 2 \times 14$$

$$\mathbf{N_{boiler\ tube} = 1,417\ no.}$$

https://doi.org/10.1515/9783110757088-066

67 Furnace total tubes weight

Furnace total tubes weight will be calculated by the following sequence:
- Furnace total tangent wall tubes weight
- Furnace membrane wall tubes weight
- Furnace front and rear wall tubes weight
- Furnace total tubes weight

67.1 Furnace total tangent wall tubes weight

$$m_{\text{tangent wall}} = N_{\text{tangent tube}} \times L_{\text{tangent tube}} \times m_{\text{tube per feet}}$$

$$L_{\text{tangent tube}} = \frac{L_{\text{max expand to steam drum}} + L_{\text{max expand to mud drum}} + H_{\text{between drums}}}{12}$$

Tab. 67.1: Carbon steel tube average weight per foot [38].

Tube outside diameter (in)	0.065 16 GA	0.085 14 GA	0.095 13 GA	0.105 12 GA	0.120 11 GA	0.135 10 GA	0.150 9 GA	0.165 8 GA
0.75	0.51339	0.6504	0.7150	0.7765	0.8650	0.9474	–	–
1	0.7031	0.8978	0.9914	1.082	1.214	1.340	1.461	1.575
1.25	–	1.145	1.268	1.388	1.563	1.733	1.897	2.055
1.5	–	1.393	1.544	1.694	1.913	2.126	2.333	2.536
2	–	1.887	2.097	2.305	2.611	2.912	3.210	3.496
2.25	–	2.135	2.374	2.610	2.960	3.305	3.643	3.976
2.5	–	2.382	2.650	2.916	3.310	3.698	4.080	4.456
3	–	–	3.203	3.527	4.010	4.483	4.953	5.417
3.25	–	–	–	3.833	5.357	4.876	5.390	5.897
3.5	–	–	–	4.138	4.706	5.269	5.826	6.380

https://doi.org/10.1515/9783110757088-067

$$L_{\text{tangent tube}} = \frac{124.16 + 13.42 + 8.38}{12} = 12.16\ \text{ft}$$

$$m_{\text{tangent wall}} = 201 \times 12.16 \times 2.305$$

$$\boldsymbol{m_{\text{tangent wall}} = 5,636\ \text{lb}}$$

67.2 Furnace membrane wall tubes weight

$$m_{\text{membrane wall}} = N_{\text{furnace membrane tube}} \times L_{\text{furnace membrane tubes}} \times m_{\text{tube per feet}}$$

$$m_{\text{membrane wall}} = 110 \times 33.6 \times 2.305$$

$$\boldsymbol{m_{\text{membrane wall}} = 8,520\ \text{lb}}$$

67.3 Furnace front and rear wall tubes weight

$$m_{\text{furnace front\&rear wall}} = 2 \times N_{\text{furnace Width tangent tube}} \times H_{\text{furnace}} \times m_{\text{tube per feet}}$$

$$m_{\text{furnace front\&rear wall}} = 2 \times 51 \times \frac{144 \times (1 + 10\%\ \text{margin})}{12} \times 2.305$$

$$\boldsymbol{m_{\text{furnace front \&rear wall}} = 3,108\ \text{lb}}$$

67.4 Furnace total tubes weight

$$m_{\text{furnace total tube}} = m_{\text{tangent wall}} + m_{\text{membrane wall}} + m_{\text{furnace front\& rear wall}}$$

$$m_{\text{furnace total tube}} = 5,636 + 8,520 + 3,108$$

$$\boldsymbol{m_{\text{furnace total tube}} = 17,264\ \text{lb} = 7,831\ \text{kg}}$$

68 Front and rear wall header weight

$$m_{\text{front and rear wall header}} = \frac{2 \times \pi \times \left(D^2 - (D - 2 \times t)^2\right) \times L_{\text{furnace width}} \times \rho_{\text{header}}}{144 \times 12} \times \left(1 + \frac{\%\text{Miscellaneous}}{100}\right)$$

$$m_{\text{front and rear wall header}} = \frac{2 \times \pi \times \left(6^2 - (6 - 2 \times 0.28)^2\right) \times 100 \times 490.059}{144 \times 12} \times (1 + 0.25)$$

$$\boldsymbol{m_{\text{front and rear wall header}} = 356.55\ \text{lb} = 161.73\ \text{kg}}$$

https://doi.org/10.1515/9783110757088-068

69 Superheater package weight

Superheater package weight will be calculated by the following sequence:
- Superheater tubes weight
- Superheater inlet header weight
- Superheater outlet header weight
- Superheater package weight

69.1 Superheater tubes weight

$$m_{\text{superheater tubes}} = m_{\text{tube per feet}} \times N_{\text{tube row}} \times N_{\text{tube row deep}} \times L_{\text{superheater tube}}$$

Tab. 69.1: Carbon steel tube average weight per foot [38].

Tube outside diameter (in)	0.065 16 GA	0.085 14 GA	0.095 13 GA	0.105 12 GA	0.120 11 GA	0.135 10 GA	0.150 9 GA	0.165 8 GA
0.75	0.51339	0.6504	0.7150	0.7765	0.8650	0.9474	–	–
1	0.7031	0.8978	0.9914	1.082	1.214	1.340	1.461	1.575
1.25	–	1.145	1.268	1.388	1.563	1.733	1.897	2.055
1.5	–	1.393	1.544	1.694	1.913	2.126	2.333	2.536
2	–	1.887	2.097	2.305	2.611	2.912	3.210	3.496
2.25	–	2.135	2.374	2.610	2.960	3.305	3.643	3.976
2.5	–	2.382	2.650	2.916	3.310	3.698	4.080	4.456
3	–	–	3.203	3.527	4.010	4.483	4.953	5.417
3.25	–	–	–	3.833	5.357	4.876	5.390	5.897
3.5	–	–	–	4.138	4.706	5.269	5.826	6.380

$$m_{\text{superheater tubes}} = 1.913 \times 19 \times 4 \times 40 \times \frac{7,600}{7,850}$$

$$\boldsymbol{m_{\text{superheater tubes}} = 5,630.31\,\text{lb}}$$

https://doi.org/10.1515/9783110757088-069

69.2 Superheater inlet header weight

$$m_{\text{superheater inlet header}} = \frac{\pi \times \left(D^2 - (D - 2 \times t)^2\right) \times L_{\text{inlet header}} \times \rho_{\text{header}}}{144 \times 12}$$

$$m_{\text{superheater inlet header}} = \frac{\pi \times \left(6^2 - (6 - 2 \times 0.344)^2\right) \times \left(\frac{19 \times 4.03125}{12} + 2\right) \times 490.059}{144 \times 12}$$

$$\mathbf{m_{\text{superheater inlet header}} = 174.29\ lb}$$

69.3 Superheater outlet header weight

$$m_{\text{superheater outlet header}} = \frac{\pi \times \left(D^2 - (D - 2 \times t)^2\right) \times L_{\text{outlet header}} \times \rho_{\text{header}}}{144 \times 12}$$

$$m_{\text{superheater outlet header}} = \frac{\pi \times \left(8^2 - (8 - 2 \times 0.5)^2\right) \times \left(\frac{19 \times 4.03125}{12} + 2 + 3\right) \times 474.453}{144 \times 12}$$

$$\mathbf{m_{\text{superheater outlet header}} = 441.61\ lb}$$

69.4 Superheater package weight

$$m_{\text{superheater package}} = m_{\text{superheater tubes}} + \left(m_{\text{superheater inlet header}} + m_{\text{superheater outlet header}}\right) \times \left(1 + \frac{\%\text{margin}}{100}\right)$$

$$m_{\text{superheater package}} = 5{,}630.31 + (174.29 + 441.61) \times \left(1 + \frac{10}{100}\right)$$

$$\mathbf{m_{\text{superheater package}} = 6{,}342.66\ lb = 2{,}876.98\ kg}$$

70 Steam drum to superheater connection header weight

$$m_{\text{connecting drum to superheater}} = \frac{\pi \times \left(D^2 - (D - 2 \times t)^2\right) \times L_{\text{connecting drum to superheater}} \times \rho_{\text{pipe}}}{144 \times 12}$$

$$m_{\text{connecting drum to superheater}} = \frac{\pi \times \left(6^2 - (6 - 2 \times 0.28)^2\right) \times 77.6 \times 490.059}{144 \times 12}$$

$$\boldsymbol{m_{\text{connecting drum to superheater}} = 1,328.1\ \text{lb} = 602.42\ \text{kg}}$$

https://doi.org/10.1515/9783110757088-070

71 Bank tube package weight

$$m_{\text{bank tube packge}} = N_{\text{total bank tubes}} \times m_{\text{tube per feet}} \times L_{\text{average bank tube}}$$

Tab. 71.1: Carbon steel tube average weight per foot [38].

Tube outside diameter (in)	0.065 16 GA	0.085 14 GA	0.095 13 GA	0.105 12 GA	0.120 11 GA	0.135 10 GA	0.150 9 GA	0.165 8 GA
0.75	0.51339	0.6504	0.7150	0.7765	0.8650	0.9474	–	–
1	0.7031	0.8978	0.9914	1.082	1.214	1.340	1.461	1.575
1.25	–	1.145	1.268	1.388	1.563	1.733	1.897	2.055
1.5	–	1.393	1.544	1.694	1.913	2.126	2.333	2.536
2	–	1.887	2.097	2.305	2.611	2.912	3.210	3.496
2.25	–	2.135	2.374	2.610	2.960	3.305	3.643	3.976
2.5	–	2.382	2.650	2.916	3.310	3.698	4.080	4.456
3	–	–	3.203	3.527	4.010	4.483	4.953	5.417
3.25	–	–	–	3.833	5.357	4.876	5.390	5.897
3.5	–	–	–	4.138	4.706	5.269	5.826	6.380
4	–	–	–	4.750	5.405	6.055	6.700	7.338

$$m_{\text{bank tube packge}} = 976 \times 2.305 \times \frac{135.07}{12}$$

$$\boldsymbol{m_{\text{bank tube packge}} = 25,322.3\,\text{lb} = 11,486\,\text{kg}}$$

https://doi.org/10.1515/9783110757088-071

72 Economizer package weight

Economizer package weight will be calculated by the following sequence:

- Economizer fin plates weight
- Economizer fin tubes weight
- Economizer return elbows weight
- Economizer headers weight
- Economizer pressure parts weight
- Economizer package weight

72.1 Economizer fin plates weight

$$m_{\text{fin plates}} = N_{\text{row no.}} \times N_{\text{row deep no.}} \times N_{\text{each tube's fin plates}} \times \rho_{\text{fin}} \times t_{\text{fin}} \times \frac{\pi \times \left((D + H_{\text{fin}})^2 - D^2 \right)}{4}$$

↓

$$N_{\text{each tube's fin plates}} = n_{\text{fin}} \times L_{\text{fin tube}}$$

$$N_{\text{each tube's fin plates}} = 6 \times 10 \times 12$$

$$\mathbf{N_{\text{each tube's fin plates}} = 720\ no.}$$

↓

$$m_{\text{fin plates}} = 10 \times 12 \times 720 \times \frac{7,750 \times 0.062428}{1,728} \times 0.04 \times \frac{\pi \times \left((2 + 0.75)^2 - 2^2 \right)}{4}$$

$$\mathbf{m_{\text{fin plates}} = 2,706.05\ lb}$$

72.2 Economizer fin tubes weight

$$m_{\text{economizer tubes}} = N_{\text{row no.}} \times N_{\text{row deep no.}} \times m_{\text{tube per feet}} \times (L_{\text{tube}} + L_{\text{tube head\& bottom connection extra}})$$

https://doi.org/10.1515/9783110757088-072

Tab. 72.1: Carbon steel tube average weight per foot [38].

Tube outside diameter (in.)	0.065 16 GA	0.085 14 GA	0.095 13 GA	0.105 12 GA	0.120 11 GA	0.135 10 GA	0.150 9 GA	0.165 8 GA
0.75	0.51339	0.6504	0.7150	0.7765	0.8650	0.9474	–	–
1	0.7031	0.8978	0.9914	1.082	1.214	1.340	1.461	1.575
1.25	–	1.145	1.268	1.388	1.563	1.733	1.897	2.055
1.5	–	1.393	1.544	1.694	1.913	2.126	2.333	2.536
2	–	1.887	2.097	2.305	2.611	2.912	3.210	3.496
2.25	–	2.135	2.374	2.610	2.960	3.305	3.643	3.976
2.5	–	2.382	2.650	2.916	3.310	3.698	4.080	4.456
3	–	–	3.203	3.527	4.010	4.483	4.953	5.417
3.25	–	–	–	3.833	5.357	4.876	5.390	5.897
3.5	–	–	–	4.138	4.706	5.269	5.826	6.380
4	–	–	–	4.750	5.405	6.055	6.700	7.338

$$m_{\text{economizer tubes}} = 10 \times 12 \times 1,887 \times \left(10 + \frac{3+3}{12}\right)$$

$$\mathbf{m_{\text{economizer tubes}} = 2,377.62\ lb}$$

72.3 Economizer return elbows weight

$$m_{\text{return elbows}} = (N_{\text{row no.}} - 1) \times N_{\text{row deep no.}} \times m_{\text{each elbow}}$$

$$m_{\text{return elbows}} = (10 - 1) \times 12 \times 1.3 \times 2.20462 \qquad 14$$

$$\mathbf{m_{\text{return elbows}} = 309.5\ lb}$$

72.4 Economizer headers weight

$$m_{\text{economizer header}} = \frac{2 \times \pi \times \left(D^2 - (D - 2 \times t)^2\right) \times L_{\text{economizer header}} \times \rho_{\text{header}}}{144 \times 4}$$

$$L_{\text{economizer header}} = \frac{(N_{\text{row no.}} - 1) \times S_{\text{L}} + L_{\text{header connection extra}}}{12}$$

$$L_{\text{economizer header}} = \frac{(10-1) \times 4.5 + 20}{12}$$

$$\mathbf{L_{economizer\ header} = 5.041\ ft}$$

↓

$$m_{\text{economizer header}} = \frac{2 \times \pi \times \left(3^2 - (3 - 2 \times 0.216)^2\right) \times 5.041 \times 490.059}{144 \times 4}$$

$$\mathbf{m_{economizer\ header} = 64.79\ lb}$$

72.5 Economizer pressure parts weight

$$m_{\text{economizer pressure part}} = m_{\text{fin plates}} + m_{\text{economizer tubes}} + m_{\text{return elbows}} + m_{\text{economizer header}}$$

$$m_{\text{economizer pressure part}} = 2,706.05 + 2,377.62 + 309.5 + 64.79$$

$$\mathbf{m_{economizer\ pressure\ part} = 5,457.99\ lb}$$

72.6 Economizer package weight

$$m_{\text{economizer package}} = m_{\text{economizer pressure part}} + m_{\text{economizer casing}}$$

↓

$$m_{\text{economizer casing}} = 1,500\ \text{lb} (\text{approximately})$$

↓

$$m_{\text{economizer package}} = 5,457.99 + 1,500$$

$$\mathbf{m_{economizer\ package} = 6,957.99\ lb = 3,156.10\ kg}$$

73 Stack weight

$$m_{\text{stack}} = \frac{\pi \times \left(D^2 - (D - 2 \times t)^2\right) \times H_{\text{active}} \times \rho_{\text{stack}}}{144 \times 4} \times \left(1 + \frac{\%\text{miscellaneous}}{100}\right)$$

$$m_{\text{stack}} = \frac{\pi \times \left(78.74^2 - (78.74 - 2 \times 0.7)^2\right) \times 57.74 \times 490.059}{144 \times 4} \times \left(1 + \frac{10}{100}\right)$$

$$\boldsymbol{m_{\text{stack}} = 74,152.02\ \text{lb} = 33,634.83\ \text{kg}}$$

https://doi.org/10.1515/9783110757088-073

74 Reports

- Water tube boiler layout
- Superheater package layout
- Economizer package layout
- Boiler performance sheet
- Stack calculation
- Forced draft fan calculation
- Pressure safety valve calculation
- Boiler feed water calculation
- Desuperheater calculation
- Steam drum water operating levels
- Weight calculation
- Flue gas velocity inside boiler

https://doi.org/10.1515/9783110757088-074

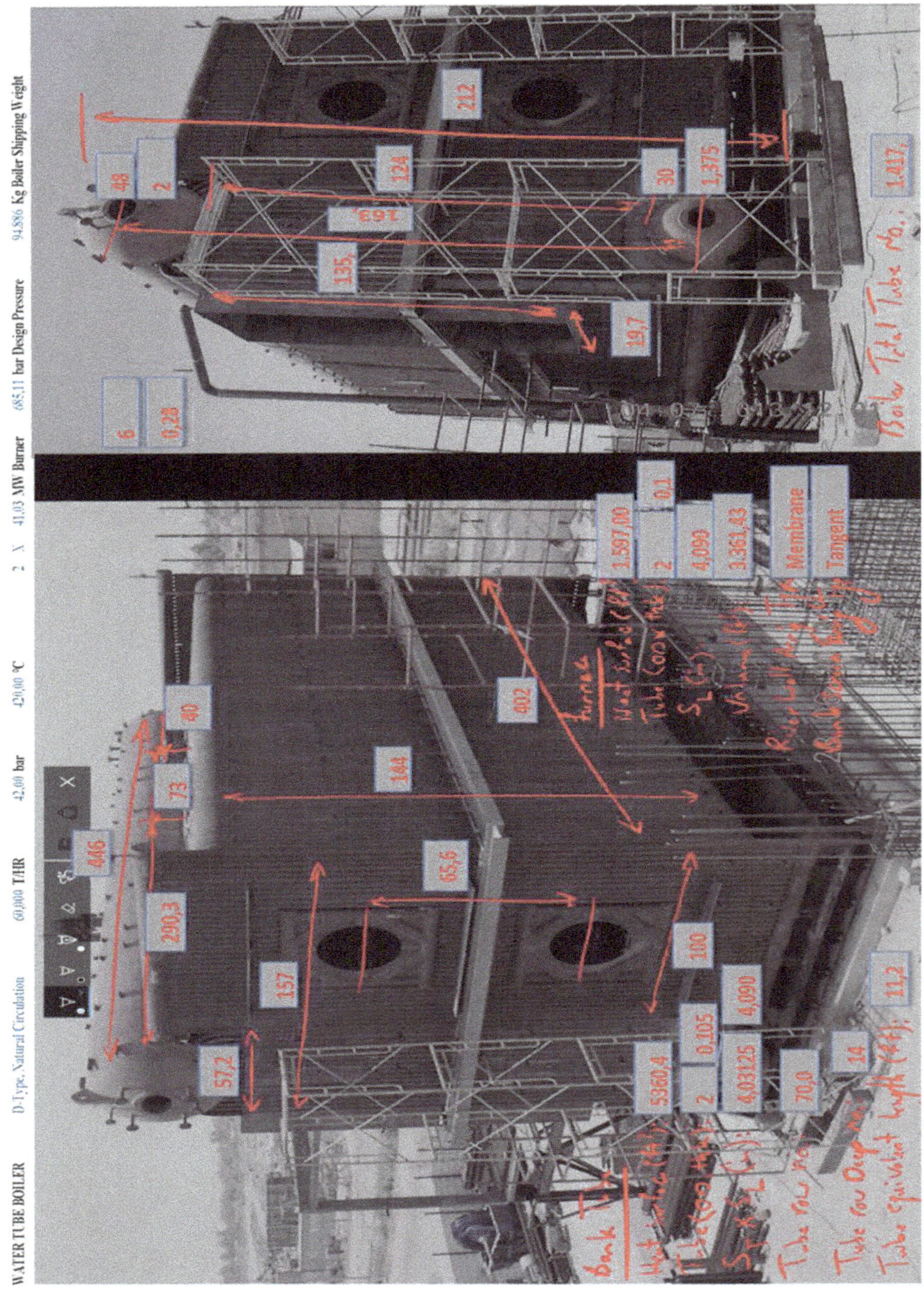

Fig. 74.1: Water tube boiler layout.

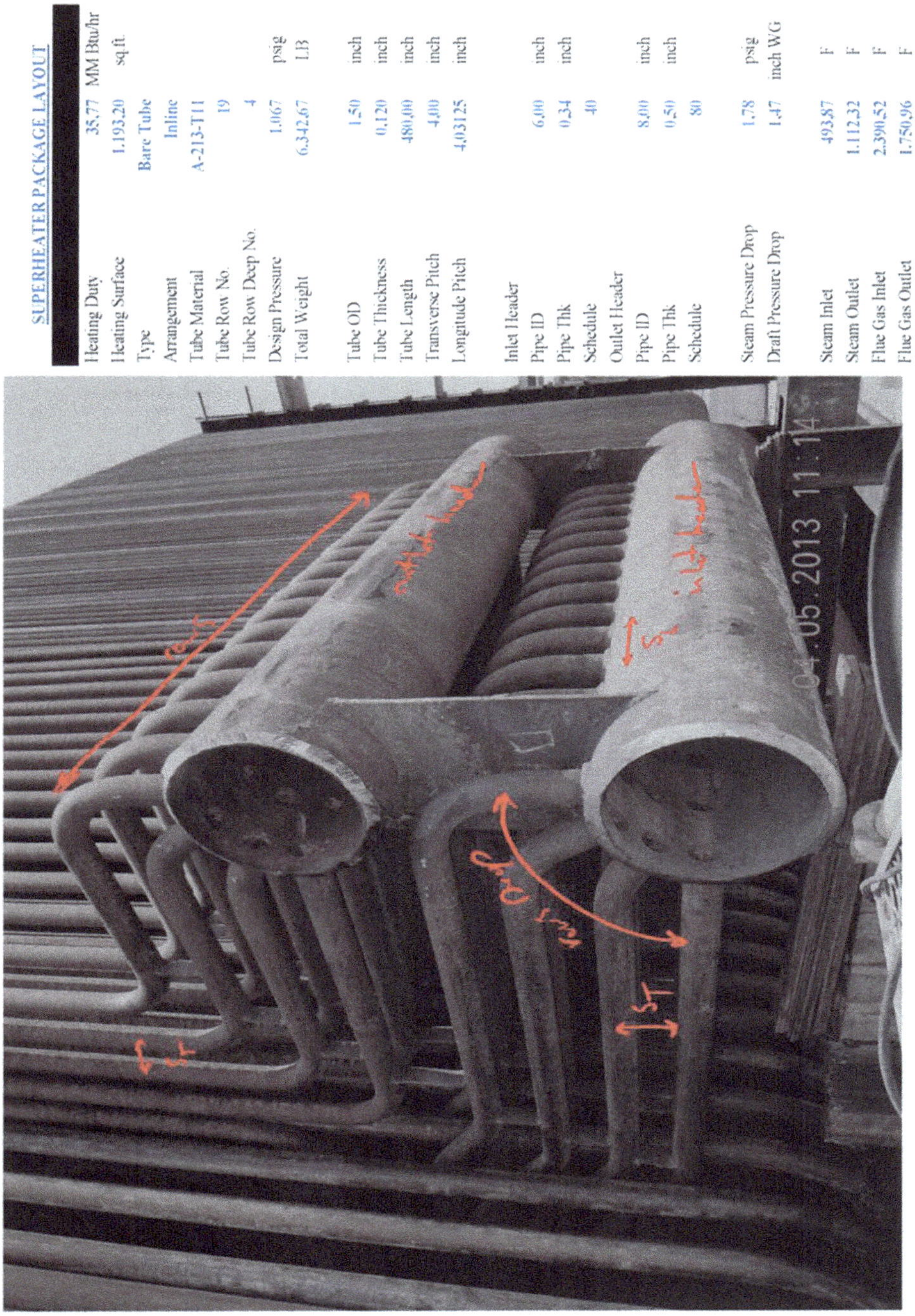

Fig. 74.2: Superheater package layout.

ECONOMIZER PACKAGE LAYOUT

Heating Duty	14.14	MM Btu/hr
Heating Surface	8,283	sq.ft.
Type	Solid	
Arrangement	Inline	
Tube Row Deep No.	12	
Tube Row No.	10	
Tube OD	2.00	inch
Tube Thickness	0.085	inch
Fin Density	6.00	Fin/inch
Fin Height	0.75	inch
Fin Thickness	0.04	inch
Fin Tube Length	120.00	inch
Transverse Pitch	4.50	inch
Longitude Pitch	4.50	inch
Inlet Header		
Pipe ID	3.00	inch
Pipe Thk	0.22	inch
Outlet Header		
Pipe ID	3.00	inch
Pipe Thk	0.22	inch
Water Pressure Drop	1.17	psig
Draft Pressure Drop	2.55	inch WG
Water Inlet	230.00	F
Water Outlet	347.14	F
Flue Gas Inlet	662.63	F
Flue Gas Outlet	368.23	F
Economizer Pressure Part Weight	5457.9929	lb
Economizer Casing Weight Approximate	1500	lb
Economizer Total Weight	6957.9929	lb
Economizer Total Weight	3156.0962	kg

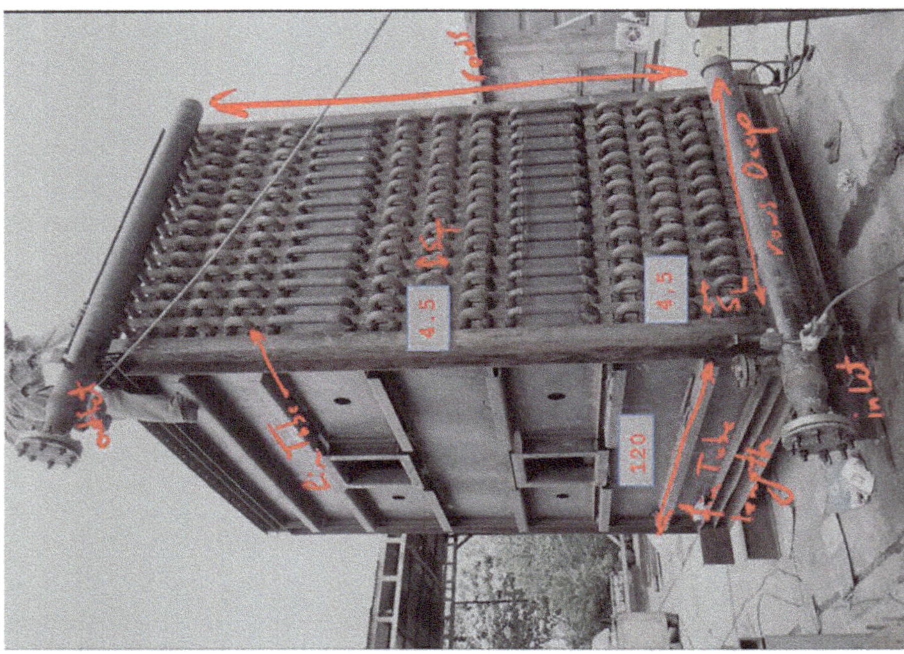

Fig. 74.3: Economizer package layout.

BOILER PERFORMANCE SHEET

MODEL			FUEL GAS		
CAPACITY	kg/hr		LHV	Kj/kg	45,933
Elevation	MASL	4	Blowdown Rate	%	3.00
Ambient air temp	°C	20	Stack Height	Meter	20.00

Steaming conditions		**100%**	**75%**	**50%**	**25%**
Steam Generation (MCR)	kg/hr	60.000	45.000	30.000	15.000
Superheater outlet pressure (SOP)	barg	42.00	42.00	42.00	42.00
Superheater pressure drop	barg	0.12	0.08	0.04	0.01
NRV valve pressure drop	barg	1.25	0.70	0.31	0.08
Steam Drum oper. pressure	barg	43.38	44.54	43.28	44.59
Feed water temperature (FWT)	°C	110	110	110	110
F.W. leaving Economizer	°C	175	170	163	157
Saturation Temperature	°C	257	258	256	258
S.H. Controlled Temp. (SOT)	°C	420	420	420	420
S.H. Uncontrolled Temp.	°C	423	423	426	424
Furnace Exit Gas Temp.	°C	1,310	1,215	1,102	908
Flue Gas Temp. Enter. S.H.	°C	1,310	1,215	1,102	908
Flue Gas Temp. Enter. Bank.	°C	955	875	780	636
Flue Gas Temp. Enter. Econ.	°C	350	318	281	239
Flue Gas Temp Leav. Econ.	°C	187	172	157	140
Flue Gas Temp Exit Of Stack	°C	178	163	148	131
Excess Air	%	17	21	28	44
Combustion Air Flow	kg/hr	72,655	56,354	39,743	22,355
Fuel Flow	kg/hr	3,835	2,876	1,918	959
Fuel Input (LHV)	KW	48,933	36,700	24,467	12,233
Flue Gas Recirc. Rate	%	0	0	0	0
Flue Gas Flow	kg/hr	77,482	60,098	42,383	23,841
Volumetric Heat Release (LHV)	Kcal/m3.hr	404,403	303,302	202,201	101,101
Surface Heat Release (LHV)	Kcal/m2.hr	283,589	212,691	141,794	70,897

Losses : Dry Gas	%	5.35	4.85	4.33	3.74
H2 & H2O in fuel	%	11.16	11.04	10.92	10.78
H2O in air	%	0.08	0.08	0.07	0.06
Radiation	%	0.50	0.56	0.66	0.89
Carbon Loss	%	0.00	0.00	0.00	0.00
Unburned Fuel Loss	%	0.00	0.00	0.00	0.00
Mfr's Margin	%	1.00	1.00	1.00	1.00
TOTAL	%	18.08	17.53	16.99	16.47
EFFICIENCY (HHV)	%	81.92	82.47	83.01	83.53
EFFICIENCY (LHV)	%	90.68	91.29	91.90	92.47

Draft Loss :					
Air Ducting	mm H2O	4.86	3.12	1.69	0.62
Inlet Damper	mm H2O	NA	NA	NA	NA
Burner Windbox	mm H2O	250.00	125.00	62.50	15.63
Furnace	mm H2O	0.201	0.078	0.040	0.013
Superheater	mm H2O	35.60	24.04	9.78	2.85
Boiler Bank	mm H2O	39.98	22.67	10.60	3.05
Economizer	mm H2O	64.85	39.30	20.10	7.05

Fig. 74.4: Boiler performance sheet.

STACK CALCULATIONS (DRAFT AND LOSSES)

Stack height	65.62	ft	Above Ground
Active length	57.74	ft	Above Flue Gas Entrance
Outlet dia.	6.56	ft	
Flue gas flow	170.818.92	lb/hr	
flue gas outlet temp	352.32	F	
Reynold No.	573.476.24		
Frict. MOODY fact.	0.011		
Elevation	13.12	ft	ASL
Out. air temp.	68.00	F	
Inside gas velocitY	1.669.61	ft/min	Typical: 2000 To 5000 Fpm
Outlet gas velocity	2.756.52	ft/min	Typical : Above 3000 Fpm
Théorical draft	0.275	in. WG	(Typical Gas; V=24.5 @ 1000 R & 30 "Hg)
Pressure losses	0.013	in. WG	(Théorical)
Net Draft (Losses)	0.262	in. WG	(Théorical)
Stack Height/I.D.=	10.00		<18 To Avoid Excessive Vibration Or Use Spoilers
Thermal Conductivity Of Insulation	0.07	Btu/ft.hr.F	
Minimum Thickness of Insulation	3.35	Inch	

Fig. 74.5: Stack calculation.

FAN CALCULATION		
Elevation	13	FASL
P atm	29,91	in.Hg
SYSTEM PRESSURE LOSS		
Inlet Screen	0,0000	in. WG
Silencer	0,0000	in. WG
Air Inlet Duct	0,1914	in. WG
Damper	0,0000	in. WG
Air Measurement	0,0000	in. WG
Air Heater	0,0000	in. WG
Air Preheater	0,0000	in. WG
Burner,Windbox	9,8425	in. WG
Boiler	0,0079	in. WG
Superheater	1,4016	in. WG
Flue Gas Duct	1,5742	in. WG
Economizer	0,0796	in. WG
Damper	2,5530	in. WG
Air Heater	0,0160	in. WG
Stack	0,0134	in. WG
Misc.	0,5000	in. WG
TOTAL	16,1797	in. WG
NET CONDITION		
Fresh Air		
Mass Flow	162.683,59	lb/hr
Volume Flow	35.492,67	SCFM
Volume Flow	36.048,20	ACFM
Temperature	68,00	°F
Density	0,08	lb/cu.ft
Minimum Temp.	-4,00	°F
Flue Gas Recir		
Fgr Mass Flow	0,00	lb/hr
Fgr Volume Flow	0,00	ACFM
Temperature	368,23	°F
Density	0,05	lb/cu.ft
Inlet Mixed		
Mixed Mass Flow	162.683,59	lb/hr
Mixed Volume Flow	36.048,20	ACFM
Temperature	68,00	°F
Density	0,08	lb/cu.ft
TEST BLOCK CONDITION		
	10	%
Static Pressure	21	%
Temperature	25	°F
Mass Flow	162.683,59	lb/hr
Volume Flow	39.653,02	ACFM
Static Pressure	19,58	in. WG
Temperature	93,00	°F
Density	0,08	lb/cu ft
Metric		
Mass Flow	73.792,12	kg/hr
Volume Flow	67.370,88	m3/hr
Static Pressure	497,27	mm H2O
Temperature	33,89	° C
Density	1,20	kg/m3
FAN'S DRIVER BHP		
Fan Efficiency	70	%
Assumed Fan Efficiency	85	%
BHP	174	BHP
BHP Cold Start-Up	183	BHP
BHP Required	205	BHP
Motor BHP	153	KW
Motor BHP (from table)	200	KW

Fig. 74.6: Forced draft fan calculation.

SAFETY VALVE SIZING

Total Steam Generation	134.400	PPH				
Superheater Operating Pressure	609,1569147	psig				
S.H. Outlet Temperature	1112,316567	F UNCONTROLLED				
Degree Superheat	618,4455669	F				
Superheater Pressure Drop	1,811351574	psi	typical; 25 to 50			
S.H. Enthalpy	1400,143132	Btu/lb				
S.H. Specific Volume	1,12930425	cu.ft/lb				
Design Pressure (Mawp)	698,6181025	psig				
NRV Delta P	13,343496	psig				
STEAM DRUM OUTLET						
Steam Flow	134.400	PPH				
Drum Operating Pressure	629,1516379	psig				
Operating Temperature	494,9042	F				
Enthalpy	1202,958948	Btu/lb				
Specific Volume	0,709043587	cu.ft/lb				
SAFETY VALVES SELECTION						
Superheater						
Relieve Capacity	39.778	PPH	SHOULD RELIEVE BETWEEN 15 & 20 %			
Corrected Sat. Flow	#REF!	PPH	Ksh	#REF!	%	
Set Pressure	**659**	psig				
Drum						
Relieve Capacity	107.520	PPH	MIN.75% OF TOTAL CAP.(PG-68.2)			
PSV-1 Setting	680	psig				
PSV-2 Setting	685	psig				
Superheater						
Kunkle 600NHG	600#	1 1/2x2 1/2 J-orifice	set @	658,81	39.778	PPH
Boiler						
Kunkle 600NJG	600#	2x3 K-orifice	set @	680,27	58.978	PPH
Kunkle 600NHG	600#	2x3 K-orifice	set @	685,11	58.978	PPH
87,76%	>75%OK					
	Total Capacity Drum Valves				117.956,00	PPH
	Total Safety Valve Capacity				157.734,00	PPH
					117,3616071	%
Superheater PSV Weight	75	lb				
Steam Drum #1 PSV Weight	95	lb				
Steam Drum #2 PSV Weight	95	lb				
Boiler Total PSV Weight	265	lb				

Fig. 74.7: Pressure safety valve calculation.

BOILER FEED WATER CALCULATION

Design Pressure	699 PSIG			
Operating Pressure	641 PSIG			
Steam Flow:	134.400 PPH			
Sat. Steam Temp.	495 F			
S.H. Steam Temp.	788 F			
Feedwater Temperature	230 F			
Blowdown Rate	5,00 %			
Feedwater Data		MIN.	NORM.	MAX.
Feedwater Flow	USGPM	57	299	356
Safety Valve Higher Setting	699 PSIG			
Min, Feedwater Inlet Pressure To Drum	719 PSIG			
Total Pressure Drop	30 PSIG			
Req'D Feedwater Pump Outlet Pressure	749 PSIG			

Fig. 74.8: Boiler feed water calculation.

DESUPERHEATER CALCULATION

Boiler Load %MCR	100%		75%	50%	25%
Steam Flow At Inlet:	134.400	PPH	100.800	67.200	33.600
Steam Flow At Outlet:	136.920	PPH	102.453	68.279	33.702
Water Flow Needed	2.520	PPH	1.653	1.079	102
Pressure Of Steam At Inlet:	609	PSIG	609	609	609
Pressure Of Steam At Outlet:	606	PSIG	606	606	606
Temperature Of Steam At Inlet:	793	°F	793	798	796
Temperature Of Steam At Outlet:	788	°F	788	788	788
Temperature Of Water	230	°F	230	230	230
Enthalpiy, Steam In	1.423	BTU/LB	1.420	1.419	1.404
Enthalpiy, Steam Out	1.400	BTU/LB	1.400	1.400	1.400
Enthalpiy, Water	200	BTU/LB	200	200	200
Degree Superheat, Steam In	298	°F	300	307	305
Degree Superheat, Steam Out	293	°F	295	296	297
Steam Flow At Outlet	136.920	PPH	102.453	68.279	33.702

Fig. 74.9: Desuperheater calculation.

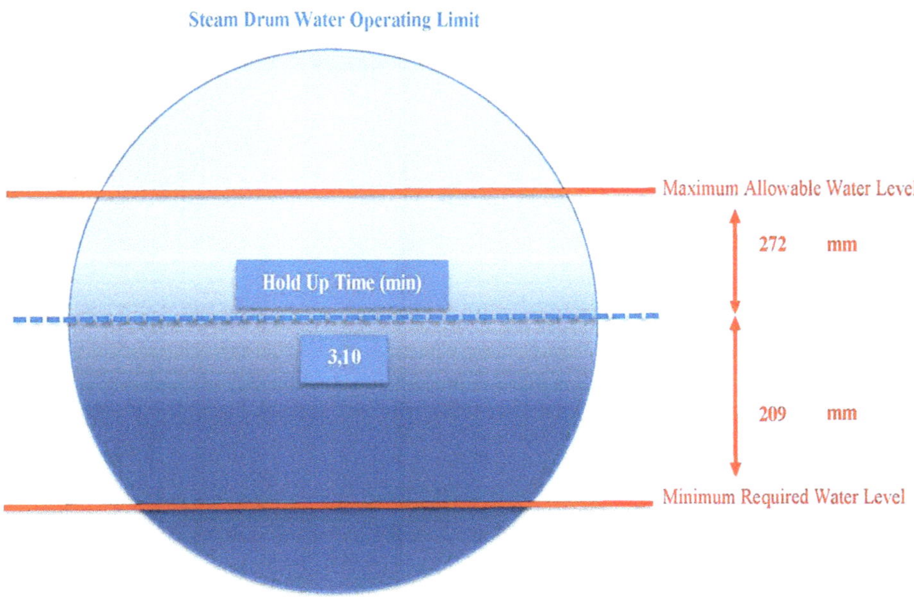

Fig. 74.10: Steam drum water operating levels.

BOILER WEIGHT

Steam Drum	19.475	kg
Mud Drum	8.125	kg
Furanace Tubes	7.831	kg
Total Front & Rear Wall Headers	162	kg
Bank Tubes	11.486	kg
Superheater	2.877	kg
Connecting Steam Drum To Superheater	602	kg
Economizer	3.156	kg
Misclenouse @ 10%	537	kg
Duct (Estimate)	7.000	kg
Stack	33.635	kg
Boiler Pressure Part Weight	53.714	kg
Boiler Shipping Weight	94.886	kg
Boiler Water Weight	17.880	kg
Boiler Operating Weight	**112.766**	**kg**

Fig. 74.11: Weight calculation.

FLUE GAS VELOCITY INSIDE BOILER

Fd.Fan Discharge Duct	3.600	fpm
Furnace	2.117	fpm
Superheater	5.297	fpm
Bank Tube	4.104	fpm
Bank Tube End	4.104	fpm
Gas Oulet Duct	4.110	fpm
after Economizer	3.046	fpm
Stack	1.670	fpm

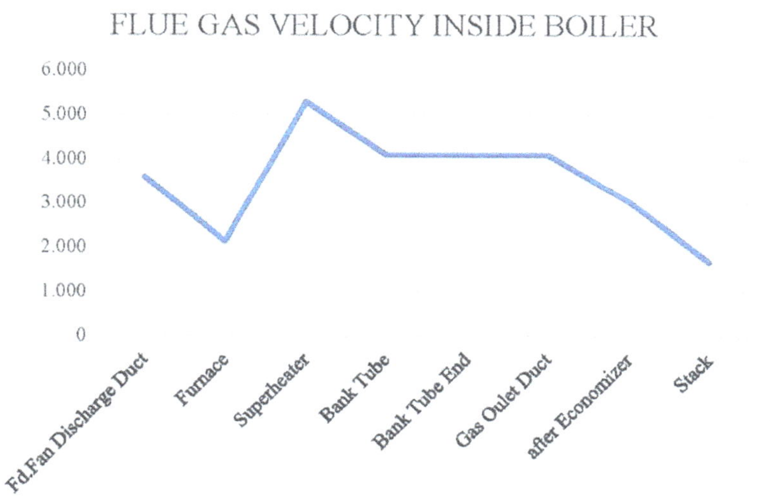

Fig. 74.12: Flue gas velocity inside boiler.

References

[1] https://en.wikipedia.org/wiki/Boiler.
[2] Characterization of the U.S. Industrial/Commercial Boiler Population by Energy and Environmental Analysis, Inc.
[3] Characterization of the U.S. Industrial/Commercial Boiler Population. Submitted By: Energy and Environmental Analysis, Inc. May 2005.
[4] https://blog.beamex.com/what-is-barometric-pressure
[5] Industrial Boilers and Heat Recovery Steam Generators Design, Applications, and Calculations. V. Ganapathy published Marcel Dekker, Inc.
[6] https://www.kane.co.uk/knowledge-centre/what-is-excess-air.
[7] https://www.pdblowers.com/tech-talk/scfm-standard-cfm-vs-acfm-actual-cfm/.
[8] https://en.wikipedia.org/wiki/Actual_cubic_feet_per_minute.
[9] Boiler Furnace Design Methods, Prabir Basu, Cen Kefa, Louis Jestin.
[10] Fundamental Laws of Physics and Calculation of Heat Transfer in Combustion Chambers of Gas-Turbine Plants. Anatoly Nikolaevich Makarov. Department of Electrical Supply and Electrical Engineering, Tver State Technical University, Tver, Russia.
[11] https://en.wikipedia.org/wiki/Reynolds_number.
[12] https://en.wikipedia.org/wiki/Prandtl_number.
[13] https://en.wikipedia.org/wiki/Nusselt_number.
[14] https://www.slideshare.net/amoso/the-reciprocal-of-thermal-conductivity-is-thermal-resistivity.
[15] https://en.wikipedia.org/wiki/Draft_(boiler).
[16] 2009 ASHRAE Handbook – Fundamentals.
[17] http://www.southwarkmetal.com/tech-tip-2-equivalent-lengths/.
[18] Prabhat S. http://www.differencebetween.net/science/difference-between-mawp-and-design-pressure.
[19] Boiler Corrosion. Zaki Ahmad, in Principles of Corrosion Engineering and Corrosion Control, 2006.
[20] ASME boiler and pressure vessel (sec.1 code PG-27).
[21] https://en.wikipedia.org/wiki/Heat_transfer_coefficient.
[22] https://www.industrialheating.com/blogs/14-industrial-heating-experts-speak-blog/post/89862-what-is-lmtd#:~:text=The%20log%20mean%20temperature%20difference,each%20end%20of%20the%20exchanger.
[23] Boiler Operation and Maintenance Manual NAVSEA created by John Pike http://www.fas.org/man/dod-101/sys/ship/eng/steam.htm.
[24] NIOEC-SP-0050 NIOEC Specification for Process and Mechanics.
[25] ASME boiler and pressure vessel code (Sec.1 code PG-52.4).
[26] Boilers for Power and Process written by Kumar Rayaprolu.
[27] https://en.wikipedia.org/wiki/Economizer.
[28] Modeling of Flow and Temperature Filed in An Economizer written by František BRUMERČÍK, Dusan SOJCAK, Michal LUKÁČ, Aleksander NIEOCZYM, Sławomir WIERZBICKI.
[29] Understanding Finned Heat Exchangers written by Viswanathan Ganapathy.
[30] http://gazogenerator.com/boilers-for-power-and-process/heat-recovery-steam-generators/.
[31] Crane, Appendix A, A26, Physical Properties of Fluids and Flow Characteristics of Valve, Fitting and Pipes.
[32] Unit-14 Design of Steel Chimneys, IGNOU Publisher, Block-3 Steel Structures. http://egyankosh.ac.in//handle/123456789/31117.

https://doi.org/10.1515/9783110757088-075

[33] Fan efficiency formula – Perbanas Institute Jakarta https://old.perbanas.id›curtains-l-tip›fan-efficiency.

[34] WEG AUSTRALIA PTY LTD, HGF Electric Motor Catalog, BRO_MO_HGFMtrTechnicalCatalogue_BROA017_0611_3K.

[35] ASME Sec. I, Power Boiler Safety Valves, PG 68.

[36] TYCO Flow Control, 2009, Kunkle Safety and Relief Products, Models 300 and 600, KUKMC-0387-US-0906.

[37] Tangent or Membrane Type Wall Tube. https://www.brighthubengineering.com/.

[38] Average Weight Per Foot Carbon Steel Boiler Tube, Condenser and Heat Exchanger Tubes SA-178 GRADE A / SA-214. anderson_tube_inventory Anderson Tube Co., Inc.

[39] Ultra-low nox burners catalog (Bulletin 4211).

[40] https://www.slawinski.co.uk/products/ellipsoidal-heads/.

[41] Steam Its Generation and Use by The Babcock & Wilcox Company (a MCDermott Company) Edition 41 Edited by J.B. Kitto and S.C. Stultz.

Index

https://doi.org/10.1515/9783110757088-076

www.ingramcontent.com/pod-product-compliance
Ingram Content Group UK Ltd.
Pitfield, Milton Keynes, MK11 3LW, UK
UKHW050842101026
15624UKWH00018B/441